Chromatography in Biotechnology

ACS SYMPOSIUM SERIES **529**

Chromatography in Biotechnology

Csaba Horváth, EDITOR
Yale University

Leslie S. Ettre, EDITOR
Yale University

Developed from two symposia sponsored
by the Division of Analytical Chemistry
of the American Chemical Society
at the Fourth Chemical Congress of North America
(202nd National Meeting of the American Chemical Society),
New York, New York,
August 25–30, 1991

American Chemical Society, Washington, DC 1993

Library of Congress Cataloging-in-Publication Data

Chromatography in biotechnology / Csaba Horváth, editor; Leslie S. Ettre, editor

    p.   cm.—(ACS symposium series, ISSN 0097–6156; 529.)

    "Developed from a symposium sponsored by the Division of Analytical Chemistry of the American Chemical Society at the Fourth Chemical Congress of North America (202nd National Meeting of the American Chemical Society) New York, NY, August 25–30, 1991."

    Includes bibliographical references and index.

    ISBN 0–8412–2669–5

    1. Chromatographic analysis—Congresses.   2. Biotechnology—Congresses.

    I. Horváth, Csaba, 1930– . II. Ettre, Leslie S.   III. Series.

TP248.25.S47C47   1993
660'.6'028—dc20   93–16982
                                              CIP

The paper used in this publication meets the minimum requirements of American National Standard for Information Sciences—Permanence of Paper for Printed Library Materials, ANSI Z39.48–1984. ∞

TP248
.25
S47
C47
1993
CHEM

Foreword

THE ACS SYMPOSIUM SERIES was first published in 1974 to provide a mechanism for publishing symposia quickly in book form. The purpose of this series is to publish comprehensive books developed from symposia, which are usually "snapshots in time" of the current research being done on a topic, plus some review material on the topic. For this reason, it is necessary that the papers be published as quickly as possible.

Before a symposium-based book is put under contract, the proposed table of contents is reviewed for appropriateness to the topic and for comprehensiveness of the collection. Some papers are excluded at this point, and others are added to round out the scope of the volume. In addition, a draft of each paper is peer-reviewed prior to final acceptance or rejection. This anonymous review process is supervised by the organizer(s) of the symposium, who become the editor(s) of the book. The authors then revise their papers according to the recommendations of both the reviewers and the editors, prepare camera-ready copy, and submit the final papers to the editors, who check that all necessary revisions have been made.

As a rule, only original research papers and original review papers are included in the volumes. Verbatim reproductions of previously published papers are not accepted.

M. Joan Comstock
Series Editor

Contents

Preface ix

1. Chromatographic Separations in Biotechnology 1
   John Frenz

   NOVEL OPERATIONAL MODES IN PREPARATIVE
   CHROMATOGRAPHY

2. Continuous Purification of Proteins by Selective
   Nonadsorptive Preparative Chromatography 14
   T. K. Nadler and F. E. Regnier

3. Ion-Exchange Displacement Chromatography of Proteins:
   Theoretical and Experimental Studies 27
   Steven M. Cramer and Clayton A. Brooks

4. Process Chromatography in Production of Recombinant
   Products 43
   Walter F. Prouty

5. Preparative Reversed-Phase Sample Displacement
   Chromatography of Peptides 59
   R. S. Hodges, T. W. L. Burke, A. J. Mendonca,
   and C. T. Mant

6. Displacement: Chromatographic Concentration Control 77
   Jana Jacobson

   CHROMATOGRAPHY OF GLYCOCONJUGATES

7. Quantitative Monosaccharide Analysis of Glycoproteins:
   High-Performance Liquid Chromatography 86
   R. Reid Townsend

8. Separation of Glucose Oxidase Isozymes from *Penicillium
   amagasakiense* by Ion-Exchange Chromatography 102
   Henryk M. Kalisz and Rolf D. Schmid

9. Analysis of Microsomal Cytochrome P-450 Patterns: Fast Protein Liquid Chromatography with Ion-Exchange and Immobilized Metal Affinity Stationary Phases....................... 112
   P. H. Roos

10. Monosaccharide Compositional Analysis of *Haemophilus influenzae* Type b Conjugate Vaccine: Method for In-Process Analysis... 132
   Charlotte C. Yu Ip and William J. Miller

ADVANCES IN COLUMN ENGINEERING

11. Zirconium Oxide Based Supports for Biochromatographic Applications 146
   P. W. Carr, J. A. Blackwell, T. P. Weber, W. A. Schafer, and M. P. Rigney

12. Preparative Reversed-Phase Chromatography of Proteins.......... 165
   Geoffrey B. Cox

Author Index 183

Affiliation Index 183

Subject Index... 183

Preface

RECENT ADVANCES IN CHROMATOGRAPHY are closely associated with biotechnology's challenge to provide efficient means for the isolation, purification, and analysis of bioproducts. Because of its versatility, chromatography has met this challenge, and, as a result, chromatographic techniques and processes play a pervasive role in all levels of biotechnology. A comprehensive treatment of chromatography in biotechnology would require a multivolume treatise. This book can cover only a few segments of this fast-growing field.

Most of this book is devoted to preparative and process chromatography of proteins and peptides. Industrial-scale purification processes require a departure from the linear elution mode of chromatography and the introduction of other separation schemes. Although many of the processes themselves are not new, their applications in biotechnology are. Often, progress in chromatography has been driven by the introduction of column materials that exhibit thermodynamically and kinetically superior properties. It is therefore appropriate that some of the chapters in this book deal with recent advances along these lines.

The use of high-performance liquid chromatography as a multifarious instrumental technique is increasing in analytical biotechnology despite the recent emergence of capillary electrophoresis. Because many new therapeutic protein products are glycosylated, efficient methods for the analysis of complex carbohydrates are increasingly in demand. The chapters on the chromatography of glycoconjugates reflect the importance of this area of research and account for some of the new developments. This is a fertile field, and we can expect further significant advances in the chromatography of complex carbohydrates.

CSABA HORVÁTH
LESLIE S. ETTRE
Department of Chemical Engineering
Yale University
New Haven, CT 06520

December 16, 1992

Chapter 1

Chromatographic Separations in Biotechnology

John Frenz

Genentech, Inc., 460 Point San Bruno Boulevard,
South San Francisco, CA 94080

This review describes the role of chromatography in the development of recombinant protein pharmaceuticals. The versatility and efficacy of chromatographic techniques have made them the *sine qua non* in both large scale and analytical separations in biotechnology. The principal qualities of chromatographic approaches that underlie this prominent role and the main techniques employed at both production and analytical scales are discussed.

A new era has dawned in technology and commerce with the advent of the new biology by which complex protein molecules can be expressed at large scale in living cells for production of human therapeutic pharmaceuticals. The developments in molecular biology and biochemistry that underlie these production processes have been revolutionary in facilitating the production of large quantities of compounds that once were scarce in pure form. The sources of these compounds, relatively fast-growing cells altered to express the target protein, may differ markedly from their natural source. Hence, in two important ways--scale and origin--the recombinant DNA-based production methods are fundamentally distinct from the traditional approaches to the preparation of biological extracts for therapeutic use. The new manufacturing techniques thus pose challenges both to the protein pharmaceutical industry and the regulatory bodies established to police the industry (*1*). These challenges for the pharmaceutical industry stem from the relative newness of the technology that limits the experience available for the design and optimization of manufacturing processes. The technology for achieving high purity separations at large scale and for analytically characterizing the resulting product have been developed to meet the unique demands of the industry. The challenges to regulatory bodies also stem from the novelty of this class of therapeutics that differ both from small molecule drugs that are the products of organic chemical syntheses, and from blood and tissue extracts that are the traditional sources of therapeutic proteins. The purification of recombinant protein pharmaceuticals requires the separation of the product from host cell proteins and DNA and from components of the culture medium. Along with these contaminants are the stringent requirements for removal of endotoxins from products expressed in E. coli and viral particles from mammalian cell culture produced proteins. In rising to these challenges the biotechnology industry has broken ground in demonstrating that proteins can be produced and recovered at scales large enough to treat common diseases, and at a purity level unprecedented for biological products (*2*).

0097–6156/93/0529–0001$06.00/0

The advances in recombinant DNA technology that gave rise to what is recognized now as the biotechnology industry began with the discovery of techniques for isolating the genes for individual proteins, expressing these proteins in microbial or mammalian cell culture and recovering the heterologous protein in active form. The allure of these techniques is that the cells are fast growing relative to whole animals and can be engineered to produce high levels of the desired protein. Thus recombinant DNA technology provides a more practical means of production of many human proteins for the treatment of disease than do conventional methods for extraction from blood and body parts. The commercial success of recombinant protein pharmaceuticals is suggested by the sales figures shown in Table I. The implementation of these techniques for the production of therapeutic proteins for human use, however, has required the development of a myriad of new technologies and new approaches to pharmaceutical manufacturing. Recombinant proteins are most conveniently expressed in bacteria, a very fast growing microbe that is relatively simple to stably transfect with the target gene. A large number of proteins, however, can only be correctly expressed in mammalian cells, either because they are incorrectly folded in the reducing environment inside bacteria (3) or require post-translational modification such as glycosylation to maintain activity in clinical use (4). Thus methods for stable transfection of continuous mammalian cell lines and their growth in large scale cultures have been developed (5). A number of methods, depending on the compartment in which the host cell deposits the protein, are employed for initial recovery of the protein from the cell broth (6). For example, many proteins expressed at high levels in bacteria form inclusion bodies that can be recovered from the lysed cell by centrifugation. Other proteins in bacteria may be secreted, as are products produced in mammalian cells, so either centrifugation or filtration may be used to separate cell mass from the product. The crude material recovered by these methods is then resolubilized, in the case of an inclusion body-bound protein (7), prior to entering the process purification stream. Final purification is generally a multi-step procedure involving complementary unit operations--often predominantly chromatographic--that give a final product in an acceptable yield, cost and purity level. Since each protein has its own distinct chemical, physical and biochemical properties, the selection and ordering of unit operations to solve the final purification problem for each product is unique. The individuality of each protein also means that the assays for purity and characterization of each product are tailored to that molecule. Analytical techniques are required to guide the selection of manufacturing conditions, from the fermentor through to the development of the final purification train. To supply these needs an impressive array of high performance tools--again, predominantly chromatographic--has been developed for the analytical protein chemist (8). Within their individual disciplines, process development and analytical chemists have their own specific constraints in the selection of their tools, and in particular in their approach to implementation of chromatographic separations.

TABLE I. BIOTECHNOLOGY SUCCESSES
World-wide sales figures adapted from reference 54.

Product	Annual Sales (Billions)
Insulin	$1.0
Growth hormone	0.9
Interferons	0.5
Erythropoietin	0.4
Tissue plasminogen activator	0.2

Large scale chromatographic operations

Among the essential considerations involved in the design of processes for the large scale purification of pharmaceuticals is the adherence of the process to regulatory guidelines. The regulation of the use of protein therapeutics derived from natural sources has been oriented toward the assurance of a consistent manufacturing process (9). This viewpoint was adopted to recognize the heterogeneous and relatively ill-defined composition of products extracted from biological sources. A drug deemed safe and effective in controlled clinical trials was assumed to remain safe and effective so long as the manufacturing route taken to produce the drug was invariant. Hence the manufacturing process, rather than detailed chemical characterization, defined the product. Regulatory bodies maintain this perspective for biologicals produced by recombinant DNA technology, so that manufacture of genetically engineered proteins is also tightly controlled, and changes in the production process require clinical testing and prior approval before the drug made by the altered process can be sold. This philosophy informs many of the decisions taken in the development of the production process for a protein pharmaceutical. Since the manufacturing process must be fixed well before marketing is approved, and often relatively early in clinical testing, important technical decisions must be made while the need for the product is still relatively low. The consequences of these decisions, however, may persist into the end stages of clinical testing or beyond, after the drug is approved and full market scale achieved. Thus, in many cases there may be a premium placed on rapidly developing the process to provide material for initial clinical trials, while ensuring the resulting process is suitable to meet future needs should the drug continue to show promise. Chromatographic separations are of singular importance to the biotechnology industry because they deliver high purity, are relatively easy to develop at the laboratory scale and can be scaled linearly to the desired production level in most cases. Hence one reason for the ubiquity of chromatographic steps in preparative protein purification steps is that they provide a relatively efficient means to meet the major manufacturing goals of the pharmaceutical biotechnology industry. The importance of this unit operation therefore accounts for the attention focused on greater understanding of the behavior of proteins in chromatographic separations, through a combination of empirical study, modeling and laboratory investigations of the adsorption of proteins on surfaces and of alternative modes of operating chromatographic columns.

One of the principal features distinguishing preparative from analytical scale chromatography is the composition and morphology of the packing material in the column (10). In interactive modes of chromatography, such as ion exchange, reversed phase or hydrophobic interaction, the packing material anchors the stationary phase (11). In modes of chromatography such as size exclusion, the packing material is inert to interaction with the feed components, and serves instead to sieve the mixture. Hence the chemical characteristics of the support material are important to its efficacy. The packing must either possess the appropriate chemical groups for its intended use or be readily modified to attach the desired chemical moieties at the surface. The packing must also be stable to buffers and cleaning reagents employed in the purification process (10). The physical strength of the support also figures importantly in its suitability to chromatographic applications since the packed bed must be able to withstand the pressure drop associated with liquid flow without deleterious deformation, collapse or fracturing of the resin. A variety of materials have been introduced to meet these needs, including synthetic polymer (12-14), polysaccharide (15) and silica-based (11) packings. The polymeric resins have the important advantages of being stable to washing with alkaline cleaning solutions that are widely used for aseptic storage of equipment, while silica is mechanically stronger and cheaper to manufacture. The morphology of the resin is controlled by the manufacturing procedure, and is an essential characteristic determining its utility for protein separations (16). The size range of the packing particles and of the internal porosity of the resin contribute to both the efficiency and capacity of the chromatographic column. Smaller diameter resins yield increased efficiency at the expense of higher pressure drop across the column, but in any size range a narrow range of particle diameters maximizes the column efficiency. The pore size and size distribution control the extent to which the resin is accessible to components of the feed mixture, and so controls the selectivity afforded by size

exclusion chromatography and the capacity of the column in interactive modes of
chromatography. In an extreme case, the pores can be wide enough to allow convective
flow through the particle that can improve the resolution afforded by large scale HPLC
purification steps (17).

The microstructure of the support determines the efficiency and capacity of the
column, but in interactive modes of chromatography the surface characteristics of the
resin control the selectivity of the separation. One of the attractions of chromatographic
approaches to protein purification is the variety of resins available that offer a broad
array of complementary selectivities (18). Table II illustrates the use of different
combinations of steps to purify enzymes from microbial sources. The principal classes
of columns include hydrophobic interaction, reversed phase, cation exchange and anion
exchange media. Hydrophobic interaction chromatography (HIC) exploits the
interaction between a protein and relatively hydrophobic stationary phase that is
promoted by increasing the salt content of the mobile phase (19). Reversed phase
chromatography also separates according to the hydrophobicity of the protein, but
typically employs harsher conditions, including more hydrophobic columns and the use
of organic solvents to elute the feed components, that can alter the conformation of the
protein (20). Ion exchange chromatography resolves proteins according to the degree of
interaction with surface-bound negatively charged (cation exchange) or positively
charged (anion exchange) moieties (21). The strength of the electrostatic interaction
between the protein and the resin surface is a function of eluent pH and ionic strength, so
these are the principal variables manipulated to effect binding to and elution from the
column. A protein can present both anionic and cationic faces to the column surface
under a given set of conditions, and so the binding to a column may be less influenced
by the net charge of the whole protein than by the charge at particular sites that are
favored for interaction with a given column. Hence, a protein may bind to both anion
and cation exchangers under a given set of conditions (22). Mixed mode resins, such as

TABLE II. PURIFICATION OF ENZYMES FROM MICROBIAL SOURCES
Adapted from reference 55

Enzyme	Purification step	Protein (mg/ml)	Specific activity (units/mg)	Recovery (%)
Aspartase from E. coli	Ammonium sulfate	27	0.4	100
	DE-52 column	2.4	4.1	120
	AX1000 column	0.4	58	84
Thiolase from C. acetobutylicum	Crude extract	1090	3.1	100
	Ammonium sulfate	530	5.8	89
	DEAE-Sephacel column	32	57	54
	Blue Sepharose column	7	111	23
	AX300 column	3	220	20
Phosphotrans-butylase from C. acetobutylicum	Crude extract	4235	7.2	100
	Ammonium sulfate	840	28.3	78
	Phenyl Sepharose column	102	182	61
	DEAE-Sephacel column	22	694	50
	AX100 column	6.2	1460	30

hydroxyapatite columns (23), that combine both anionic and cationic exchange functionalities also can provide altered useful selectivities compared to monophyletic exchangers. In addition, interactions of varying degrees of specificity have been exploited in affinity chromatographic separations (24). Affinity separations can be carried out with a wide variety of immobilized ligands, from metals to enzyme substrate analogs to monoclonal antibodies, that are chosen to specifically interact with the desired protein. In many cases this approach simplifies the purification process. With all types of columns, the selectivity can vary significantly according to the precise method of manufacturing a particular column. The manufacturing method can have a qualitative influence on the capacity or selectivity of the resin, so optimization of a separation scheme necessarily includes screening a variety of packings for enhanced performance.

The column defines the selectivity available for a given separation, but the capacity and throughput of the process is also determined by the mode of operation of the column. Among the choices made in defining the process are the steps by which the feed components are bound and eluted. The choice of operating mode in preparative purification operations has an effect on the capacity of the chromatographic step. Since the cost of a column step increases with the size of the column, operating modes have been investigated that maximize the capacity of the column while maintaining the resolution required for the desired purity level. Figure 1 shows schematically the different operating steps employed in chromatography. The simplest operating mode is isocratic elution. In this mode the column is equilibrated and eluted with a buffer adjusted so that the feed components transit the column at different velocities, separating as they move down the column. The operational simplicity of this approach is that no changes in buffer composition are made at the column inlet. In size exclusion chromatography this is the most common operating approach. However in many types of interactive chromatography of proteins it can be difficult to reproducibly achieve the precise buffer composition that yields the appropriate degree of interaction to allow isocratic elution or it may be impossible to find buffer conditions in which all components of the mixture are eluted from the column. Hence, in most separations, as in analytical chromatography of proteins, a gradient in eluent strength is employed to elute the column (25). The change in eluent strength may be continuous or stepwise, according to the degree of resolution required. Continuous gradients, including linear gradients, in eluent strength require relatively sophisticated

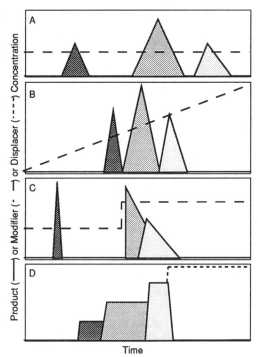

Figure 1. Schematic chromatograms for various operating modes in liquid chromatography, including (A) isocratic elution, (B) gradient elution, (C) step elution, and (D) displacement. The dashed lines represent (A-C) modifier and (D) displacer concentrations.

pumping, mixing and control equipment to deliver reproducible eluent composition profiles to the column. This approach is more practical, however, than isocratic elution for biopolymers, whose binding to stationary phases can be reversed with a small change in eluent strength (26). In order to simplify the operational mechanics of gradient separations, stepwise elution protocols are commonly employed. In these methods, the feed mixture is bound to the resin under conditions in which, preferably, some impurities flow through the column. The buffer composition is then changed in a stepwise fashion to elute the desired product while retaining on the column certain other impurities. Generally, stepwise elution is simpler to implement at large scale provided binding and elution conditions can be found that provide the necessary resolution. In certain cases, it may be more efficient to implement a related operating mode, frontal chromatography, in which the binding conditions are selected so that the desired product is unretained and the column binds impurities. The advantage of this approach is that none of the binding capacity of the column is taken up by the desired product, instead the column is sized for impurity removal. In the later stages of a process, where the feed stream is relatively pure, frontal chromatography thus offers a much more efficient use of the resin. Another operationally simple method is displacement chromatography (28-30), in which the feed mixture is bound to the column, and rather than be eluted, the product is displaced from the resin by a stepwise input of a solution of a compound that has a high affinity for the surface. Physically, the difference between displacement and stepwise elution mechanisms is that the displacer is more strongly retained than the feed mixture, while an eluent is less strongly retained. In practice, in displacement chromatography the displacer zone remains behind the product bands while the eluent in elution chromatography is mixed in with the product bands. Displacement separations yield purified product bands that attain a thermodynamically fixed concentration, with sharp boundaries between adjacent bands in the effluent concentration profile. This is in contrast to stepwise elution techniques, that result in relatively concentrated, but tailing, peaks in the chromatogram. The displacement mode can have advantages in capacity, resolution and processing simplicity compared to elution techniques, provided that the relatively narrow constraints on displacer properties can be satisfied.

Protein analysis by HPLC

In part, the approach to regulating the manufacture of pharmaceuticals derived from biological sources described above stemmed from the relative scarcity of purified blood- and tissue-derived products that effectively precluded in-depth chemical analysis. The low abundance of therapeutically active proteins in the complex matrices that are the classical sources of these proteins impeded large scale manufacture to extremely high purity and the degree of analytical characterization that was routine for synthetic small molecule drugs. The products of the biotechnology revolution are, however, available in relatively large quantities so not only is the production of high purity protein pharmaceuticals possible, but also the detailed characterization of the resulting drug. One aspect of confirming the consistency of the manufacturing process has thus become the confirmation of the consistency of the resulting product by analytical protein chemistry techniques (1). The revolution in biotechnology has therefore been accompanied by a revolution in analytical techniques, some of which are shown in Figure 2, for the detailed description of protein structures (31).

The degree of complexity of the interplay of biological and biochemical processes involved in the fermentation and purification of a protein pharmaceutical is such that analytical characterization provides important evidence supporting the consistency of the process or can provide the basis for troubleshooting and improving a process. Analytical scale chromatographic techniques--especially HPLC--are essential tools for providing the data to meet these objectives. Hence the last decade has seen an explosion in the breadth of applications of chromatography to analytical biotechnology and in the availability of columns and chromatographic systems tailored to these

Analytical technique	Property					
	Subunit structure	Conformation	Primary sequence	Post-translational modifications	Degradation	Homogeneity
Spectroscopic						
Optical	•	•		•		
MS	•		•	•	•	
NMR	•	•		•		
X-ray crystallography	•	•		•		
Separations						
Electrophoresis	•			•	•	•
HPLC	•	•	•	•	•	•
Chemical						
Amino acid composition			•	•	•	
N-terminal sequencing	•		•	•	•	
CHO composition				•		
Enzymatic						
C-terminal sequencing			•			
Peptide mapping			•	•	•	
CHO structure				•		

Figure 2. Analytical tools employed for different aspects of the structural characterization of proteins. Adapted from reference *31*.

applications. One area in which chromatography has provided insights that are not available by any other technique is in the characterization of carbohydrate structures on glycoproteins (*32*). These structures can be essential to proper folding and activity of the protein, and can strongly influence the pharmacokinetics, bioavailability and efficacy of a protein pharmaceutical, yet are usually sufficiently heterogeneous to present a daunting analytical challenge (*33*). Modern analytical approaches nevertheless have been developed that separate and quantify the constituent monosaccharides or the individual glycoforms arrayed on the protein. This level of description of a glycoprotein is a testament to the advances recently made in the field of analytical biotechnology that confers a new level of meaning to the widely-coined phrase "high performance". The performance evinced by these applications clearly represents a new height.

The achievement that gave birth to HPLC and underlies its continuing importance as the preeminent method in analytical protein chemistry was the adoption of micron-sized particles as supports for column packing materials (*34*). The use of these materials entailed the development of costly high pressure columns and equipment in order to effect separations, but dramatically improved the speed, resolution and sensitivity of separations compared to those available by conventional low pressure LC approaches. Continuing refinements and growth in the technology of HPLC has produced the current broad spectrum of columns and equipment to meet the needs of the equally diverse assortment of HPLC users (*35*). The variety of analytical columns and the high efficiency of analytical HPLC columns together contribute to the approach taken by analytical chemists to separations problems that is markedly distinct from that taken by designers of large scale protein purification operations. As described above, large scale separations typically involve several columns with different selectivities that together yield high purity. Analytical separations, however, typically employ only a single column so much more reliance is placed on high efficiency of the column to provide the resolution required to ascertain the purity and composition of the sample. This distinction accounts for the widespread adoption of HPLC for analytical separations and the drive over the last two decades to increase the efficiencies of columns and instrumentation. This drive has expanded to account in large part for the allure of capillary electrophoresis that has demonstrated efficiencies in certain applications that far outpace even HPLC.

Despite the improvements in instrument and column technologies that have steadily improved the efficiencies of HPLC separations, in many instances the physico-chemical behavior and structural heterogeneity of proteins preclude high efficiency separations by techniques that perform well for small molecule separations. Hence, renewed attention has focused, as it has in large scale separations, on exploiting the differences that are manifest among the rich variety of HPLC column packing materials that are commercially available. Traditionally, the reversed phase mode has dominated-- and almost become synonymous with--HPLC. The denaturing conditions associated with the reversed phase mode, however, tend to mitigate the differences among proteins in a manner not unlike the denaturing conditions of SDS-PAGE (36). This mitigation has the effect, on the one hand, of narrowing the peaks obtained in protein separations, but, on the other hand denaturation can obliterate the structural differences among proteins that provides the most useful basis for separation (37). Thus in many cases analytical separations, like preparative chromatography, must be carried out in aqueous buffers in the size exclusion, hydrophobic interaction, ion exchange or various affinity modes of operation. Each of these modes of chromatography separates according to a different property of the analyte and so provides a different basis for selectivity of a given analytical separation. In addition, many chemically different columns suited to each of these modes are available from various commercial sources. The differences among columns manufactured by different routes can be sufficient to significantly alter the selectivity obtained in a separation, particularly of proteins, that interact in a myriad of subtle ways with biological and synthetic surfaces (38). Therefore, an important element of the full power of HPLC for protein separations derives from the plethora of columns available that differ in large and small ways from one another.

Another aspect of HPLC that confers much of its power as an analytical tool is the variety of methods of detection that can enhance the sensitivity of the approach as well as provide on-line characterization data on the separated components of a mixture (39). The workhorse detection method is UV-absorbance, which is convenient and relatively sensitive for compounds such as peptides and proteins. Scanning and diode-array detectors expand the utility of UV detection by providing an absorbance spectrum that yields additional information on the identity of a peak (40). Certain compounds, including tryptophan-containing proteins, fluoresce under UV-illumination, and so fluorescence detectors can provide higher sensitivity and selectivity than absorbance detection, particularly in the analysis of proteins from complex samples such as serum or cell culture broths. Electrochemical detection is also selective and so is a powerful approach for the quantitative identification of compounds that contain an electrochemically labile group. One important application of electrochemical detection is the analysis of carbohydrates released chemically or enzymatically from a glycoprotein, using pulsed amperometric detection (PAD) (32). Since carbohydrates are UV-transparent, PAD methods provide a high sensitivity alternative to derivatization techniques for chromatographic detection. Other detection methods employed in HPLC provide limited characterization data that can be interpreted to identify the peaks eluted from the chromatographic column. Light-scattering detection has been employed to determine the molecular size of monomeric and aggregated proteins in a variety of modes of HPLC (*41*). Detection by on-line mass spectrometry provides molecular mass information that can be sufficient to identify a protein or peptide following HPLC (*42*). The recent development of the electrospray ionization interface (*43*) to the mass spectrometer has revolutionized MS and LC-MS of peptides and proteins by extending both the sensitivity and mass range of high resolution instruments. The broadening acceptance of mass spectrometry in the life sciences, particularly following the commercialization of the electrospray interface, has underscored the utility of MS as a tool of analytical protein chemistry and the symbiosis of HPLC and MS in many biochemical applications. The detection methods available for HPLC continue to make advances in technology to improve, in significant ways, the sensitivity, selectivity and information content provided in applications involving proteins. These improvements have already served to entrench even further the prominent role of HPLC in the protein chemistry laboratory.

One of the most important applications of HPLC in analytical biotechnology is peptide mapping: the separation of the mixture of peptides produced by enzymatic digestion of a protein (*44*). The combination of the proteolytic selectivity afforded by digestion with proteases and the high resolution obtained with HPLC yields peptide maps that are unrivaled for identification and primary structure confirmation of a protein (*45*). "Restriction proteases" such as trypsin cleave a protein into peptide fragments whose small size makes them much better behaved in HPLC systems than, usually, is the intact protein. The small size of the constituent peptides is also usually more amenable to rapid characterization by a combination of amino acid analysis, mass spectrometry and N-terminal sequencing than is the entire protein in one piece. Furthermore, enzymatic mapping procedures permit the ready identification of sites of glycosylation (*45*) and disulfide bond formation (*46*) or of other post-translational modifications that are not apparent from the cDNA sequence of a protein. The ubiquity of enzymatic mapping approaches to protein characterization can thus be traced to the close match between the capabilities of standard techniques employed in protein chemistry and the optimal analyte size for highest resolution in HPLC. Advances in mass spectrometer and protein sequencer design are only just beginning to enlarge the potential for detailed characterization of the sequence of intact proteins, so peptide mapping can be expected to remain a standard tool in the arsenal of the protein chemist for some time to come.

A second important and rapidly evolving area in which HPLC separations have become essential biochemical tools is the characterization of the carbohydrate structures found on the surfaces of glycoproteins (*47*). The growing usage of recombinant glycoproteins as pharmaceuticals has intensified the appreciation of the role of glycosylation in mediating the structural integrity, physicochemical, biochemical and biological activity of a protein as well as its pharmacokinetic disposition *in vivo*. The appreciation of the impact of glycosylation on so many properties of a protein has induced an awareness on the part of manufacturers of the need to monitor these attributes in the recombinant product. Figure 3 shows the biosynthetic processing of oligosaccharides that goes on in the rough endoplasmic reticulum (RER) and Golgi organelles, and illustrates some of the diversity of structures that can result. HPLC approaches have made a large impact in carbohydrate characterization efforts, in particular the combination of high pH anion exchange chromatography and pulsed amperometric detection (HPAE-PAD) (*48*). These instruments allow either mapping of the oligosaccharides liberated by endoglycosidase digestion of a glycoprotein (*49*) or composition profiling of the monosaccharides released by acidic hydrolysis (*50*). In both cases anion exchange chromatography at high pH permits high resolution separations of closely related carbohydrate species, and PAD is a sensitive, efficient means of detecting the achromophoric sugars. The HPAE-PAD approach can also be scaled to provide preparative amounts of oligosaccharides for further structural characterization by MS, tandem MS or nuclear magnetic resonance analyses (*51*). The development of HPAE-PAD simplified the characterization of carbohydrates compared to earlier techniques, many of which relied on derivatization of the reducing end of the glycan to facilitate separation and detection. Since oligosaccharides are relatively easy to derivatize in this manner, these approaches remain important in many applications and have been the basis for powerful separations implemented in capillary electrophoresis systems (52). Finally, the separation of classes of glycoforms of intact proteins have been demonstrated in ion exchange (Frenz, J.; Quan, C.P.; Cacia, J.; MacNerny, T.; and Bridenbaugh, R. In preparation.) and immobilized metal affinity chromatographic systems (53). These approaches provide an even more rapid means of quantifying the glycosylation pattern of a protein compared to digestion methods, and permit the isolation of homeoglycosylated proteins that can be useful for investigating the influence of glycosylation on protein activity and other properties. These applications demonstrate the heights attained

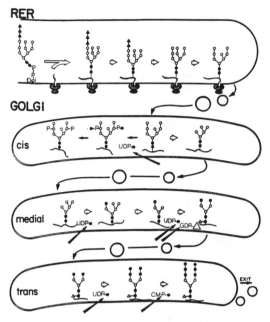

Figure 3. Schematic pathway of oligosaccharide processing on newly synthesized glycoproteins in the rough endoplasmic reticulum (RER) and Golgi bodies. The symbols represent: P, phosphate; ■, N-acetylglucosamine; O, mannose; ▲, glucose; △ , fucose; ●, galactose; ◆, sialic acid. Reprinted with permission from reference 33.

by HPLC in discriminating among subtle variations of the heterogeneous ensemble of glycoforms constituting a purified glycoprotein and should become increasingly important in the higher performance of liquid chromatography in the future.

Conclusions

This volume brings together papers illustrating the advances made in chromatographic techniques for producing proteins at large scale and for the characterization of glycoproteins. The body of papers thus illustrates the approaches to streamline the development process for large scale protein purification, and the high resolution of the analytical techniques that can be brought to bear to characterize the product. At either scale the resolution afforded by chromatographic techniques is the major factor that drives the choice of separation method. The process development chromatographer deploys his arsenal of columns in order to select a single protein in high yield from among the thousands of compounds that may be present in a cell culture broth. The analytical chromatographer similarly has his armory to tackle the problems of resolving and defining the contributors to microheterogeneity in a purified product. One shared goal is the demonstration of manufacturing consistency, that is a prerequisite to product approval. Meeting this goal, along with the other objectives related to the development of the fruits of biotechnology, requires all the innovations in tools and techniques that continue to spring from equipment manufacturers and end-users, including those described in this volume.

Literature cited

1. Liu, D.T.-Y. *Trends Biotechnol.* **1992**, *10*, 364.
2. Garnick, R.L.; Solli, N.J.; and Papa, P.A. *Anal. Chem.* **1988**, *60*, 2546.
3. Goff, S.A.; and Goldberg, A.L. *Cell* **1983**, *41*, 587.
4. Pennica, D.; Holmes, W.E.; Kohr, W.J.; Harkins, R.N.; Vehar, G.A.; Ward, C.A.; Bennett, W.F.; Yelverton, E.; Seeburg, P.H.; Heyneker, H.L.; Goeddel, D.V.; and Collen, D. *Nature* **1983**, *301*, 214.
5. McCormick, G.; Trahey, M.; Innis, M.; Dieckmann, B.; and Ringold, G. *Mol. Cell. Biol.* **1984**, *4*, 166.
6. Cull, M.; and McHenry, C.S. In *Guide to Protein Purification*, M.P. Deutscher, Ed., Methods in Enzymology 182, Academic Press: San Diego, California, 1990; pp. 147-153.
7. Marston, F.A.O.; and Hartley, D.L. In *Guide to Protein Purification*, M.P. Deutscher, Ed., Methods in Enzymology 182, Academic Press: San Diego, California, 1990; pp. 264-276.
8. Hancock, W.S. In *High Performance Liquid Chromatography in Biotechnology*, W.S. Hancock, Ed., John Wiley and Sons: New York, 1990; p. 1.
9. Manohar, V., and Hoffman, T. *Trends Biotechnol.* **1992**, *10*, 305.
10. Low, D. In *High Performance Liquid Chromatography in Biotechnology*, W.S. Hancock, Ed., John Wiley and Sons: New York, 1990; p. 117.
11. Unger, K.K.; Lork, K.D.; and Wirth, H.-J. In *HPLC of Peptides, Proteins and Polynucleotides: Contemporary Topics and Applications*, M.T.W. Hearn, Ed., VCH Publishers: New York, 1991; p. 59.
12. Ugelstad, J.; Mork, P.C.; Kaggerud, K.H.; Ellingsen, T.; and Berge, A. *Adv. Colloid Interface Sci.* **1980**, *13*, 101.
13. Kao, Y.; Nakamura, K.; and Hashimoto, T. *J. Chromatogr.* **1983** *266*, 358.
14. Vratny, P.; Mikes, O.; Strop, P.; Coupek, J.; Rexova-Bendova, L.; and Chadimova, D. *J. Chromatogr.* **1983**, *257*, 23.
15. Hjertén, S. *Arch. Biochem. Biophys.* **1962**, *99*, 466.
16. Hjertén, S. In *HPLC of Peptides, Proteins and Polynucleotides: Contemporary Topics and Applications*, M.T.W. Hearn, Ed., VCH Publishers: New York, 1991; p. 119.
17. Regnier, F.E. *Nature (London)* **1991**, *350*, 634.
18. Henry, M.P. In *High Performance Liquid Chromatography in Biotechnology*, W.S. Hancock, Ed., John Wiley and Sons: New York, 1990; p. 21.
19. Shaltiel, S.; and Er-el, Z. *Proc. Natl. Acad. Sci. USA* **1973**, *52*, 430.
20. Frenz, J.; Hancock, W.S.; Henzel, W.J.; and Horváth, Cs. In *HPLC of Biological Molecules: Methods and Applications*, K.M. Gooding and F.E. Regnier, Eds., Marcel Dekker: New York, 1990; p. 145.
21. Regnier, F.E.; and Chicz, R.M. In *HPLC of Biological Molecules: Methods and Applications*, K.M. Gooding and F.E. Regnier, Eds., Marcel Dekker: New York, 1990; p. 77.
22. El Rassi, Z.; and Horváth, Cs. *J. Chromatogr.* **1986**, *359*, 255.
23. Kawasaki, T.; Takahashi S.; and Ikeda, K. *Eur. J. Biochem.* **1985**, *152*, 361.
24. Josíc, D.; Becker, A.; and Reutter, W. In *HPLC of Peptides, Proteins and Polynucleotides: Contemporary Topics and Applications*, M.T.W. Hearn, Ed., VCH Publishers: New York, 1991;p. 469.
25. Antia, F.D.; and Horváth, Cs. *Ann. N.Y. Acad. Sci.* **1990**, *589*, 172.
26. Kunitani, M.; Johnson, D.; and Snyder, L.R. *J. Chromatogr.* **1986**, *371*, 313.
27. Liao, A.; and Horváth, Cs. *Ann. N.Y. Acad. Sci.* **1990**, *589*, 182.
28. Tiselius, A. *Ark. Kemi. Mineral Geol.* **1943**, *16A*, 1.
29. Horváth, Cs. In *The Science of Chromatography*, S. Bruner, Ed. , Elsevier: Amsterdam, 1985; p. 179.
30. Frenz, J.; and Horváth, Cs. In *HPLC--Advances and Perspectives, Volume 5*, Cs. Horváth, Ed., Academic Press: New York, 1988; p. 211.

31. Geisow, M.J. *Biotechnol.* **1991**, *9*, 921.
32. Basa, L.J.; and Spellman, M.W. *J. Chromatogr.* **1990**, *499*, 205.
33. Kornfeld, R.; and Kornfeld, S. *Ann. Rev. Biochem.* **1985**, *54*, 631.
34. Horváth, Cs.; and Lipsky, S.R. *Nature (London)* **1966**, *211*, 748.
35. Dorsey, J.G.; Foley, J.P.; Cooper, W.T.; Barford, R.A.; and Barth, H.G. *Anal. Chem.* **1992**, *64*, 353R.
36. Laemmli, U.K. *Nature (London)* **1970**, *227*, 680.
37. Oroszlan, P.; Wicar, S.; Teshima, G.; Wu, S.-L.; Hancock, W.S.; and Karger, B.L. *Anal. Chem.* **1992**, *64*, 1623.
38. Cacia, J.; Quan, C.P.; Vasser, M; Sliwkowski, M.B.; and Frenz, J. *J. Chromatogr.;* in press.
39. Yeung, E.S., Ed. *Detectors for Liquid Chromatography*, Wiley: New York, 1986.
40. Sievert, H.-J.P.; Wu, S.-L.; Chloupek, R.; and Hancock, W.S. *J. Chromatogr.* **1990**, *499*, 221.
41. Krull, I.S.; Stuting, H.H.; and Krzysko, S.C. *J. Chromatogr.* **1988**, *442*, 29.
42. Caprioli, R.M.; Moore, W.T.; DaGue, B.; and Martin, M. *J. Chromatogr.* **1988**, *443*, 335.
43. Fenn, J.B.; Mann, M.; Meng, C.K.; Wong, S.F.; and Whitehouse, C.M. *Science* **1989**, *246*, 64.
44. Hancock, W.S.; Bishop, C.A.; and Hearn, M.T.W. *Anal. Biochem.* **1979**, *89*, 203.
45. Chloupek, R.C.; Harris, R.J.; Leonard, C.K.; Keck, R.G.; Keyt, B.A.; Spellman, M.W.; Jones, A.J.S.; and Hancock, W.S. *J. Chromatogr.* **1989**, *463*, 375.
46. Becker, G.W.; Tackitt, P.M.; Bromer, W.W.; Lefeber, D.S.; and Riggin, R.M. *Biotechnol. Appl. Biochem.* **1988**, *10*, 325.
47. Parekh, R.B.; and Patel, T.P. *Trends Biotech.* **1992**, *10*, 276.
48. Hughes, S.; and Johnson, D.C. *Anal. Chim. Acta* **1981**, *132*, 11.
49. Maley, F.; Trimble, R.B.; Tarentino, A.L.; and Plummer, T.H. *Anal. Biochem.* **1989**, *180*, 195.
50. Edge, A.S.B.; Faltynek, C.R.; Hof, L.; Reichert, L.E.; and Weber, P. *Anal. Biochem.* **1981**, *118*, 131.
51. Spellman, M.W.; Basa, L.J.; Leonard, C.K.; Chakel, J.A.; O'Connor, J.V.; Wilson, S.; and van Halbeek, H. *J. Biol. Chem.* **1989**, *264*, 14100.
52. Nashabeh, W.; and El Rassi, Z. *J. Chromatogr.* **1991**, *536*, 31.
53. Porath, J. *J. Chromatogr.* **1988**, *443*, 3.
54. Hancock, W.S. *LC-GC* **1992**, *10*, 96.
55. Rudolph, F.B.; Wiesenborn, D.; Greenhut, J.; and Harrison, M.L. In *HPLC of Biological Molecules: Methods and Applications*, K.M. Gooding and F.E. Regnier, Eds., Marcel Dekker: New York, 1990; p. 668.

RECEIVED December 15, 1992

Novel Operational Modes in Preparative Chromatography

Chapter 2

Continuous Purification of Proteins by Selective Nonadsorptive Preparative Chromatography

T. K. Nadler and F. E. Regnier

Department of Chemistry, School of Science, Purdue University, West Lafayette, IN 47907

Continuous forms of chromatography are important to the biotechnology industry for the production of therapeutic proteins. Selective non-adsorption preparative (SNAP) chromatography may be used in a cross-current, simulated moving bed to produce a continuous purification system. It has the advantages of speed, efficient use of packing material, scalability, decrease cost of pumping systems, gentleness with regard to protein denaturation, and reduced dilution of the sample. An immobilized metal affinity chromatography (IMAC) column may be added in tandem with SNAP to concentrate as well as purify the protein.

There is a great need in the biotechnology industry for separation technology of increased flexibility, specific throughput, and economic efficiency that may be used in the production of therapeutic proteins. These attributes are often associated with continuous processes and is the reason that continuous purification systems are now of great interest. Methods for continuous purification of proteins by chromatographic, electrophoretic, aqueous two-phase extraction and reversed micelle separation technology are currently being explored. This paper focuses briefly on continuous chromatographic techniques in general and primarily on a new continuous chromatographic technique called selective non-adsorption preparative chromatography (SNAP).

Continuous Forms of Chromatography

Continuous chromatographic separations were first widely used in the petroleum industry (1,2) where work initially concentrated on the development of large-scale batch separations. Continuous systems became favored because they were more flexible, could be run unattended, decrease the need for recycling effluent, and utilized sorbent materials more effectively. Much of this work has been applied recently in the development of continuous chromatographic methods for proteins (3-5).

0097–6156/93/0529–0014$06.00/0

Multiple approaches have been taken in the construction of continuous chromatography systems. Counter-current flow, cross-current flow, co-current flow, continuous stirred tanks, and selective non-adsorption are all examples of continuous systems.

Counter-Current Flow Separations. These systems are based on the mobile phase and sorbent moving in opposite directions as the name implies. Because counter-current systems generally separate a mixture into only two fractions, selectivity is very important with complex mixtures. Moving beds in which both the sorbent and solvent are transported (1,2), simulated moving beds in which the sorbent in a fixed bed appears to move through the use of valves along the column axis (6,7), a moving column approach that uses column switching to simulate sorbent transport (8,9), and fluidized bed chromatography (10,11) are all forms of counter-current separations. The problem with these systems relative to proteins is that they are generally isocratic elution systems and proteins are not easily separated isocratically.

Cross-Current Flow Separations. Annular chromatography is the most typical example of a cross-current flow system (12-16). Rotating annular columns, in which the sorbent is placed between two concentric cylinders to form an annular column, have been effective in size-exclusion separations of proteins (5). Moving column systems with a rotating bundle of columns are another version of this approach (13,17,18). Liquid in the moving column system is delivered to the columns and eluent to the collection flasks by liquid distributors. At the present time only isocratic separations of proteins have been achieved with these systems. It does not appear that the specific throughput of these systems is any higher than that of the individual columns comprising the system.

Moving Belt Separation. A very unique system has been constructed by using a moving mylar belt on which nylon pouches of immunosorbent were attached (19). The pouches resembled tea bags of immunosorbent through which proteins could diffuse without loosing the adsorbent. The belt was first dipped into the association chamber where it selectively bound the antigen of interest (in this case alkaline phosphatase). As the belt exited each chamber, the pouches were squeezed by a roller to remove excess liquid. Unbound protein was removed and antigen collected by sequential passage through a wash chamber and dissociation chamber respectively. Subsequent passage through another wash chamber recycled the sorbent pouches. A system using 5 g of immunosorbent was capable of at least a 10 fold purification with 90% recovery and a 35-40 hr cycle time (19). Although continuous, the specific throughput of this system was very low. It is questionable whether this system is easily scaled up.

Continuous Stirred Tank Reactors (CSTR). These systems have been used in situations where one or two equilibrium contact stages are sufficient for a high yield separation (3,20,21). Each stage, or contactor, is equivalent to one theoretical plate. The CSTR must have at least two stages, one for adsorption and another for desorption or recycling to be continuous. Adsorbent must be separated from the solvent in each stage. It is this separation process where CSTR systems diverge in design. The continuous affinity-recycle extraction

(CARE) system of Gordon (3) is a good example of a CSTR purification
system.   ß-galactosidase from *E. Coli* was purified 18 fold with 79%
recovery in a two stage CARE system using an affinity sorbent for
ß-galactosidase (PABTG/Agarose).   Addition of a wash stage between the
adsorbing and desorbing stage increased the purification to 22 fold and
recovery   to 90%.   In a counter-current adsorption design, the
purification jumped to 170 fold, but recovery fell to 72% (3).
Optimization of these systems is critical.   Drastic changes in yield
and recovery may be obtained with minor alterations of system design
and operation.

Selective Non-Adsorption Preparative (SNAP) Chromatography

Chromatographic separations of proteins are achieved by differential
adsorption   at   surfaces   in   all   cases   except   size-exclusion
chromatography.   These surface mediated separations are generally based
on selective elution of substances from a sorbent surface.   Elution is
achieved in either an isocratic or gradient elution regime.   Because
proteins adsorb to surfaces at multiple sites, very small changes in
solvent composition have a large impact on the adsorption/desorption
equilibrium and isocratic elution is seldom used .   Elution is more
commonly achieved with a solvent gradient of increasing desorbing
strength.   Unfortunately gradient elution is a problem in continuous
chromatography that would be desirable to circumvent.   Protein
denaturation is another problem in chromatographic systems.   Some
proteins denature when they are adsorbed to a sorbent surface (22).   It
would be desirable to purify proteins without having to adsorb them to
the chromatography column.   The fact that proteins are purified by a
series of discrete purification steps is yet another negative feature
of current separation systems.   Multiple step purifications are both
labor intensive and diminish recovery.   It would be desirable to
combine steps in protein purification.   The selective non-adsorption
preparative chromatography (SNAP) approach described below is an effort
to address these problems in the case of preparative chromatography.

     SNAP chromatography is uniquely different from other forms of
chromatography in that it removes all of the proteins in a sample
except the one of interest.   The target protein is allowed to pass
through the column unretained.   The ion-exchange form of SNAP
chromatography was first reported by Petrilli et al. for the
purification of aspartate amino-transferase (23). The system separates
proteins by taking advantage of the observation that retention of a
protein on ion-exchange chromatography columns is minimal when it is at
its isoelectric point (pI).   (It will be recalled that the net charge
of a protein is zero at its pI.)   At pH values above its pI, a protein
has a net negative charge and will adsorb to an anion exchange sorbent
(Figure 1).   In contrast, the protein is positively charged and will
adsorb to a cation exchanging matrix below its pI.   Retention of a
protein on an ion-exchange column at its pI can occur if it has an
uneven charge distribution (e.g., a group of negatively charged
residues separated from the positively charged residues).   Petrilli et
al. used an anion-exchange column followed by a cation-exchange column
(23), however, a single SNAP column may also be prepared by mixing
anion-exchange and cation-exchange resins.   The novelty of this
approach is that by operating the column at the pI of the target

protein it passes through the column unretained while other proteins
are retained.

SNAP chromatography offers several advantages over conventional
gradient-elution chromatography. (1) Since the protein of interest is
not adsorbed to the surface, its structure should not be altered by the
media. This means that the protein should retain more activity. (2)
The separation is very quick because the protein passes through the
column unretained. Loss of protein from adsorption, proteolysis, and
time dependent denaturation phenomena will be dramatically reduced.
(3) The entire chromatography column is used during the purification
process. This maximizes the efficiency of the adsorbent and increases
the amount of protein processed per volume of adsorbent. (4) SNAP
chromatography is easily scaled up for larger separations by making the
columns larger. (5) The pumping system is less expensive because
simpler isocratic systems are used in lieu of more expensive gradient
pumping systems. (6) The sample is not diluted by the separation
process because it is purified by a form of frontal chromatography.

Petrilli et al. suggested the name "Isoelectric Chromatography"
(*23*) for the ion exchange approach, however, SNAP chromatography can be
used in more than the ion-exchange mode, albeit with a different
separation mechanism. Preliminary results from Stringham et al.
(*24,25*) show that tandem hydrophobic interaction (HIC), protein A
affinity, and size-exclusion chromatography (SEC) columns may be used
in SNAP separations. SNAP separations in the affinity and immobilized
metal affinity modes may also be used. Immunoaffinity columns would be
well suited to remove trace contaminants. However, it is often
difficult to produce an antibody to each single contaminant. Some
contaminants are more immunogenic than others, so most of the
antibodies would be directed against materials of high immunogenicity
and few or no antibodies are produced against the rest. It would be
necessary to isolate each contaminant and produce antibodies against
it. This is not feasible because many of the contaminants are not even
identified, much less purified. Work reported by Anicetti, at Genetech
(*26*), involved a cascade immunization scheme to produce antibodies for
a wide range of antigens. However, their work focused on developing
antibodies for analytical methods to detect trace contaminants that
remain after preliminary purification, not the purification itself.

Description of the Continuous SNAP System. After a certain amount of
use, a SNAP chromatography column will become saturated with
contaminants and will need to be recycled. To make the process
continuous, a second column must be used while the saturated column is
being cleaned. The loading and cleaning cycle is divided into three
phases. (1) The loading phase, when sample is being purified by the
column, occupies 50% of the cycle time. (2) Desorption using a 1.0 M
salt buffer requires 25% of the cycle, and (3) re-equilibration with
the sample buffer consumes the remaining 25% of the cycle. Since
desorption and re-equilibration take no longer than loading, two
columns may be used in parallel for continuous purification. One
column is desorbed and re-equilibrated while the other is loaded. The
re-equilibrated column will be ready to load again when the other
column becomes saturated. A ten-port valve arrangement (Figure 2) was
used to operate the SNAP columns in tandem in a continuous purification

Retention Map

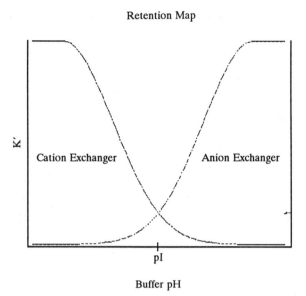

Buffer pH

Figure 1. At pH values below the pI of the protein, it is positively charged and may be adsorbed by cation exchangers, as shown by the increase in the capacity factor at low pH. At pH values above the pI of a protein, it will be negatively charged and may be adsorbed by anion exchangers.

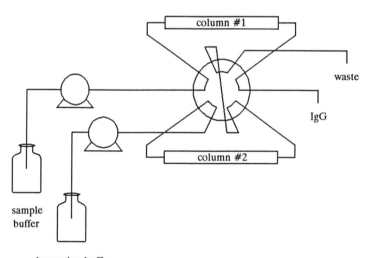

Figure 2. Dual SNAP columns were used to process a continuous stream of diluted bovine serum. A ten-port valve switched the mobile phases between serum and desorption buffer so that as one column loaded, the other was being recycled.

system.   The valve was automated with a pneumatic actuator so a computerized LC system could run the purification unattended.

The use of perfusable (POROS) packing materials for the adsorbent allows high flow rates without increasing back pressure or slowing mass transfer (*27,28*). This material has been designed with large through pores to improve mass transfer at high flow rates. Linear velocities in excess of 1 cm/sec on 250 mm x 4.6 mm columns were used in some experiments.

SNAP chromatography, as well as many other forms of continuous protein separations, functions best with relatively dilute solutions (ca. 1 mg/ml). This means that the purified protein is even more dilute because it only represented a fraction of the total protein (5% would be 50 μg/ml).   It is often necessary to concentrate the product solution by some means such as ultrafiltration.   However, the separation process itself does not dilute the sample, because the sample elutes as a frontal curve.

An immobilized metal affinity chromatography (IMAC) column was added in tandem to the SNAP columns to adsorb and concentrate the dilute product protein. IMAC (*29-34*) utilizes an immobilized metal ion to chelate histidine residues on proteins. Copper II was used because it has the broadest specificity for different proteins (*35*). In this way, the SNAP-IMAC technique is more general for proteins other than IgG.   The purified protein eluting from the SNAP columns adsorbs and concentrates on the IMAC column. The protein is then eluted from the IMAC column with a pulse of imidazole to displace the protein. Imidazole does not strip the column of the metal ion, so there is no need to reload the IMAC column with metal. Further, since imidazole is a small molecule, it may easily be removed, if necessary, from IgG by using dialysis or a size-exclusion cartridge.

Further purification, as well as concentration, may be performed on the IMAC column. By using a stepped gradient of imidazole, the contaminants that are more weakly adsorbed to the IMAC column may be eluted first with a low concentration of imidazole. Then a higher concentration of imidazole may be used to remove the concentrated protein. Finally, a very concentrated imidazole pulse could be used to remove any contaminant that remained.

The system used in these studies falls into the category of a cross-current simulated moving bed and was used to purify a single protein from a mixture. However, cross-current systems can purify more than one protein simultaneously. If gradient elution were used to desorb the saturated column, a fraction collector at the waste outlet could collect other proteins as they elute during the wash gradient. However, purification of multiple constituents simultaneously may require the addition of more valves and columns to the system. Because two or more proteins are rarely purified simultaneously, no effort was given to the purification of multiple proteins.

Observations. Concentrated feed streams load the SNAP chromatography columns faster than they may be desorbed, so the system works best with dilute feed streams (less than 1 mg/ml). If the feed stream is too

concentrated, product will also be wasted. Each time the SNAP column
is desorbed, the column still contains the components of the feed
stream in its void volume. These components include the product
protein as well as the contaminants. The product yield, or recovery of
the SNAP chromatography system is a function of the column void volume
(V_0) and the volume of feed solution that passes through the column (V_s)
before the column becomes saturated. Optimum recovery (R) is related
to these variables by the following equation:

$$R = \left(\frac{V_s}{V_s - V_0} \right) 100\%$$

To optimize recovery, V_s needs to be maximized and V_0 minimized.
Usually, very little can be done with V_0 since it is a function of the
packing material. However, the use of a packing material with a high
loading capacity does increase V_s. Also, as the feed stream is diluted,
V_s increases because it takes longer to saturate the column. On the
other hand, it takes effort to reconcentrate the product protein, so
there is a compromise between dilution and recovery.

One way to avoid the recovery problem is to use two six-port
valves instead of one ten-port valve to control the flow through the
SNAP columns. The first valve determines which column receives the
feed stream and the second valve determines which column is connected
to the collection vessel. After the first valve turns, the second
valve is delayed equivalent to the time it takes the column to pass the
void volume. Recovery could always be 100% as long as the column did
not saturate in less than one void volume.

Scale-up of the SNAP system is simply a matter of increasing
column size and flow rates. Since the separation is insensitive to
plate height, the usual hydrodynamic problems associated with larger
columns are reduced. This is an advantage over those continuous
systems that require multiple columns with equivalent flow properties.
The SNAP system only requires that the desorption time of the slowest
column be less than the loading time of the fastest column. Columns
can be matched simply by adjusting the feed rate relative to the
desorption flow rate.

In the experiments performed in these studies, standard
analytical columns (4.6 mm i.d. x 25 cm) were operated at flow rates of
10 ml/min. This means that the linear flow rate was in excess of 1
cm/second. These flow rates were possible because a perfusable
adsorbent (POROS) was used to pack the columns. High throughputs were
then possible using rapid multiple cycles to perform continuous
separations (See Table I).

The adsorbent in the column is used very efficiently in rapid
multiple cycle SNAP chromatography. The rate of protein processed for
a given volume of adsorbent may be greater than 1.0 mg/min/ml of
column.

The IMAC column was placed in tandem to the SNAP columns (Figure
3) to make a continuous purification and concentration system. An
injection valve was placed between the SNAP columns and the IMAC column

Table I. Dual Column SNAP Results

	Experiment 1	Experiment 2	Experiment 3
Flow rate (ml/min)	10	5	10
Linear Flow Rate (cm/sec)	1.0	0.5	1.0
Serum Dilution	1/100	1/25	1/50
Serum Protein Conc. (mg/ml)	0.781	2.13	1.09
Fraction IgG (rel. area 1st peak)	8.85%	6.28%˙	7.47%
Processing Rate (mg Serum/min)	7.8	10.7	10.9
Rate/Col. Volume (mg/min/ml)	0.94	1.28	1.30
IgG Protein Conc. (mg/ml)	0.084	0.061	0.047
IgG Purity (rel. area 1st peak)	~40%	> 80%	~95%
Fold Purification	4.52	12.8	12.7
Approximate Yield	48.9%	36.5%	54.8%

so pulses of a desorption buffer could be injected into the IMAC column to release the concentrated IgG. The valve was fitted with a 2 ml sample loop for that purpose. Once the IMAC column was saturated with IgG, it was eluted with 200 mM imidazole to yield 95% pure IgG at concentrations greater than 5 mg/ml. Other methods of concentration are also possible. Affinity columns for the protein may be placed after the SNAP columns to capture the dilute protein. Ion-exchange columns also may be used, but the mobile phase pH would have to be adjusted so that the protein achieved a net charge (either positive or negative). IMAC was chosen because it was simple to use (no pH adjustment was necessary), and it was fairly generic (it could be used for a number of different proteins because it is not as specific as affinity columns).

Some contaminants in the purified IgG_1 were observed by SDS-PAGE. Selective elution of these contaminants was attempted in order to obtain further purification as well as concentration from the IMAC column. Salt washes as high as 1.0 M NaCl did not elute any of the IgG nor any of the contaminants. The loaded IMAC column was removed from the SNAP-IMAC system and eluted with a gradient of imidazole. The IgG eluted from the column at a low imidazole concentration (less than 20 mM) while one contaminant eluted at 200 mM imidazole.

To take advantage of this selective elution without the use of a linear gradient, a series of pulses of increasing imidazole concentration were used. The imidazole pulse was introduced via the injection valve between the SNAP and IMAC columns. A fraction was collected after each pulse and analyzed by SDS-PAGE (Figure 4). Some high-molecular-weight contaminants are removed by a 1 mM imidazole pulse while most of the IgG elutes during the 10 mM and 15 mM pulses. A low-molecular-weight contaminant remains until a 200 mM pulse is used. By using a series of three imidazole pulses, 5 mM to elute the high-molecular-weight contaminants, 15 mM to elute the IgG and 200 mM to elute the remaining contaminant, the IgG is further purified as well as concentrated on the IMAC column.

Optimization of SNAP. Optimization of SNAP chromatography separations has been discussed in detail (24,25). To optimize the SNAP system, one must choose a salt concentration just high enough to elute the protein of interest. Higher salt concentrations compromise the separation by desorbing contaminants. To find the optimum salt concentration, Stringham et al. (25) started with an equation derived by Snyder et al. (36). The equation relates isocratic and gradient elution. Retention time (t_R) is related to average k' in gradient elution (k^*), column dead volume (t_0) and the k' at the onset of the gradient (k_0) by:

$$t_R = t_0 \ k^* \ \log\left(\frac{2.3 \ k_0}{k^*}\right) + t_0 \qquad (1)$$

The average k' is given by:

$$k^* = \frac{t_g}{\Delta\phi \ t_0 \ S} \qquad (2)$$

where t_g is the duration of the gradient, ($\Delta\phi$) is the change in volume fraction of the eluent solvent (ϕ is 1.0 when the gradient is run from 0 to 100%) and S is the change in log k' for the unit change in ϕ in isocratic elution. In isocratic systems, retention is given by:

$$\log(k') = \log(k_0) - S\phi \qquad (3)$$

Equation (2) may be reduced to:

$$k^* = \frac{t_g}{t_0 \ S} \qquad (4)$$

When the entire gradient is run from 0 to 100%. A capacity factor of 10 is in the sharp transition range discussed above. Substituting this value into equation 3 and rearranging gives:

$$\log(k_0) = 1 + S\phi \qquad (5)$$

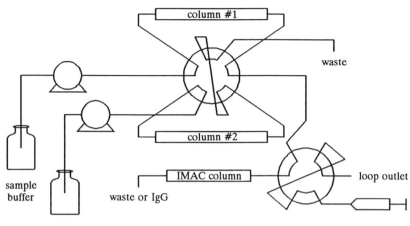

Figure 3. An IMAC column was added to the dual-column SNAP apparatus. IgG was concentrated on the IMAC column and was eluted with imidazole. A microprocessor-controlled pumping system was used to automate the system.

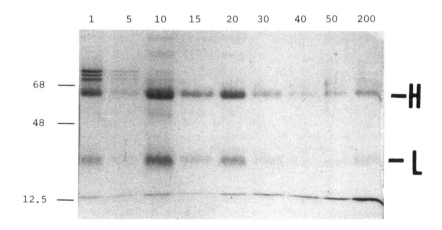

Figure 4. SDS-PAGE of samples eluted from the IMAC column at various imidazole concentrations (1 mM, 5 mM, 10 mM, 15 mM, 20 mM, 30 mM, 40 mM, 50 mM, 200 mM). IgG produces two bands because it is composed of two heavy (H) chains and two light (L) chains. Molecular weight standards are indicated on the left (68 KDa, 48 KDa, and 12.5 KDa).

Substituting (4) and (5) into equation 1 yields:

$$t_R = \frac{t_0}{t_0} \frac{t_g}{S} [\log(2.3) + 1 + S\phi - \log(\frac{t_g}{t_0 \, S})] + t_0$$

which simplifies to:

$$t_R = 1.36 \left(\frac{t_g}{S}\right) + t_g\phi - \left(\frac{t_g}{S}\right) \log\left(\frac{t_g}{t_0 \, S}\right) + t_0$$

Solving for ϕ:

$$\phi = \frac{t_R}{t_g} - \frac{t_0}{t_g} - \left(\frac{1}{S}\right) [1.362 - \log\left(\frac{t_g}{t_0 \, S}\right)]$$

Converting ϕ to salt concentration, substituting constants, A, B, and C and a k' of 10, the equation reduces to:

$$[NaCl] = A\left(\frac{t_R}{t_g}\right) - B\left(\frac{1}{t_g}\right) - C \qquad (6)$$

The constants were experimentally determined to produce the equation:

$$[NaCl] = 0.33\left(\frac{t_R}{t_g}\right) - 1.21\left(\frac{1}{t_g}\right) - 0.018$$

For a 20 min gradient on a 10 cm column, this reduces to:

$$[NaCl] = 0.0165 \, t_R - 0.078$$

Stringham et al. (25) evaluated this equation with ovalbumin and bovine serum albumin and found that the predicted salt concentration was within 0.005 M of the experimentally observed concentration. Therefore, to find the desired salt concentration for any protein, one must run a 20 min gradient for 0 to 0.5 M NaCl on a 100 x 4.6 mm mixed bed column. The experimentally determined retention time is used to solve the equation for the salt concentration needed to run the SNAP column. If the protein elutes without the use of salt, then no salt should be added to the SNAP buffer.

Stringham et al. (25) also looked at the effect of additional salt on the purification factor. They found that as the salt was increased, the purification factor decreased, but the purification factor was strongly dependent on the pH of the mobile phase and the isoelectric points of the contaminants in the sample.

Conclusions

There are many new types of continuous protein separations. Continuous SNAP chromatography is one of these new methods which utilizes adsorbent very efficiently in a dual column system. This system can be classified as a cross-current simulated moving bed. Like many other methods, the output protein concentration is low. It has been shown in this paper that coupling SNAP chromatography in tandem with IMAC solved the dilution problem by concentrating the sample.

SNAP chromatography may also be used for sample preparation in analytical separations to remove large amounts of contaminants before injection onto the analytical column. Further, SNAP chromatography is not limited to ion exchange. Any form of adsorbent may be used (eg. HIC, RPC, Affinity, IMAC etc.) in which contaminants may be selectively removed. SNAP chromatography has been applied to several proteins, not just IgG. Stringham et al. (*25*) also purified myoglobin, amylase, and carbonic anhydrase. A combination of chromatographic modes including SEC and protein A affinity have also been used in concert with the mixed bed ion exchange column.

The advantages of SNAP include simplicity of the theory, increased purification speed, gentleness on proteins, efficient use of adsorbent, simplified scale-up, and decreased cost in the pumping system. These attributes make SNAP chromatography very promising technique for the purification of proteins on both small and large scale production.

Literature Cited

1. Berg, C. *Trans. Am. Inst. Chem. Engrs.* **1946**, *42*, 665–680.
2. Kehde, H.; Fairfield, R. G.; Frank, J. C.; Zahnstecher, L. W. *Chem. Eng. Progr.* **1948**, *44(8)*, 575–582.
3. Gordon, N. F.; Tsujimura, H.; Cooney, C. L. *Bioseparation* **1990**, *1*, 9–21.
4. Mattiasson, B.; Ramstorp, M. *J. Chromatogr.* **1984**, *283*, 323–330.
5. Nicholas, R. A.; Fox, J. B., Jr. *J. Chromatogr.* **1969**, *43*, 61–65.
6. Broughton, D. B. *Chem. Eng. Progr.* **1968**, *64*, 60–65.
7. Broughton, D. B. *Sep. Sci. Technol.* **1984-85**, *19(11&12)*, 723–736.
8. Hashimoto, K.; Adachi, S.; Noujima, H.; Maruyama, H. *J. Chem. Eng. Japan* **1983**, *16(5)*, 400–406.
9. Barker, P. E.; Deeble, R. E. *Anal. Chem.* **1973**, *45*, 1121–1125.
10. Cloete, F. L. D.; Streat, M. *Nature* **1963**, *200*, 1199–1200.
11. Arehart, T. A.; Bresee, H. C.; Hancher, C. W.; Jury, S. H. *Chem. Eng. Progr.* **1956**, *52*, 353–359.
12. Svensson, S. H. *Chem. Abs.* **1952**, *46*, 4863g.
13. Svensson, H.; Agrell, C.; Dehlen, S.; Hagdahl, L. *Sci. Tools.* **1955**, *2(2)*, 17–21.
14. Fox, J. B., Jr.; Calhoun, R. C.; Eglinton, W. J. *J. Chromatogr.* **1969**, *43*, 48–54.
15. Fox, J. B., Jr. *J. Chromatogr.* **1969**, *43*, 55–60.
16. Canon, R. M.; Begovich, J. M.; Sisson, W. G. *Sep. Sci. Technol.* **1980**, *15(3)*, 655–678.
17. Dinelli, D.; Polezzo, S.; Taramasso, M. *J. Chromatogr.* **1962**, *7*, 477–484.
18. Taramasso, M.; Dinelli, D. *J. Gas Chromatogr.* **1964**, *2(5)*, 150–153.
19. Hughes, J. J.; Charm, S. E. *Biotechnol. Bioeng.* **1979**, *21*, 1439–1455.
20. Afeyan, N. B.; Gordon, N. F.; Cooney, C. L. *J. Chromatogr.* **1989**, *478*, 1–19.
21. Gordon, N. F.; Cooney, C. L. *In Protein Purification: from Molecular Mechanisms to Large-Scale Processes*; Ladisch, M. R.

        Ed.; ACS Symposium Series 427, American Chemical Society,
        Washington, DC 1990, *427*, pp 118-138.
22. Lu, X.; Figueroa, A.; Karger, B. L. *J. Am. Chem. Soc.* 1988, *110*,
        1978-1979.
23. Petrilli, P.; Sannia, G.; Marino, G. *J. Chromatogr.* 1977, *135*, 511-
        513.
24. Stringham,   R.   W.   *Selective   Non-Adsorption   Preparative
        Chromatography of Proteins*; Ph. D. Thesis, Purdue University,
        West Lafayette, IN 1989.
25. Stringham, R. W.; Grott, A. E.; Regnier, F. E. *Prep. Chromatogr.*
        1989, *1*, 179-193.
26. Anicetti, V. *In Analytical Biotechnology Capillary Electrophoresis
        and Chromatography*; Horvath, Cs.; Nikelly, J. G. Eds.; ACS
        Symposium Series 434, American Chemical Society, Washington, DC,
        1990, pp 113-126.
27. Fulton, S. P.; Afeyan, N. B.; Gordon, N. F. *J. Chromatogr.* 1991,
        547, 452-456.
28. Afeyan, N. B.; Fulton, S. P.; Regnier, F. E. *J. Chromatogr.* 1991,
        544, 267-269.
29. Porath, J.; Carlsson, J.; Olson, I.; Belfrage, G. *Nature* 1975, *258*,
        598-599.
30. Porath, J.; Olin, B. *Biochem.* 1983, *22*, 1621-1630.
31. Hemdan, E. S.; Porath, J. *J. Chromatogr.* 1985, *323*, 247-254.
32. Belew, M.; Yip, T.; Andersson, L.; Porath, J. *J. Chromatogr.* 1987,
        403, 197-206.
33. Porath, J. *J. Chromatogr.* 1988, *43*, 3-11.
34. Nakagawa, Y.; Yip, T.; Belew, M.; Porath, J. *Anal. Biochem.* 1988,
        168, 75-81.
35. Belew, M.; Yip, T. T.; Anderson, L.; Ehrnstrom, R. *Anal. Biochem.*
        1987, *164*, 457-465.
36. Snyder, L. R.; Stadalius, M. A.; Quarry, M. A. *Anal. Chem.* 1983,
        55, 1412A-1430A.

RECEIVED October 14, 1992

Chapter 3

Ion-Exchange Displacement Chromatography of Proteins

Theoretical and Experimental Studies

Steven M. Cramer and Clayton A. Brooks

Bioseparations Research Center, Howard P. Isermann Department of Chemical Engineering, Rensselaer Polytechnic Institute, Troy, NY 12180–3590

In this chapter we present a steric mass-action (SMA) ion-exchange equilibrium formalism which explicitly accounts for the steric hindrance of salt counter-ions upon protein binding in multicomponent equilibria. A simple method is presented for the rapid calculation of effluent profiles of displaced proteins and induced salt gradients under ideal chromatographic conditions. Theoretical predictions are compared to experimental results in both cation and anion exchange systems. Both the induced salt gradients and the displacement profiles match well with the theoretical predictions. These displacements differ from the traditional vision of displacement in several important ways: the isotherm of the displacer does not lie above the feed component isotherms; the concentrations of displaced proteins exceed the concentration of the displacer; and the induced salt gradient produces different salt microenvironments for each displaced protein. These results demonstrate the ability of the SMA formalism to predict complex behavior present in ion-exchange protein displacement systems. Furthermore, the development of a rapid method for obtaining ideal isotachic displacement profiles with this formalism facilitates methods development and optimization of ion-exchange protein displacement separations.

High-performance ion-exchange chromatography is widely employed for the purification of proteins in the pharmaceutical and biotechnological industries (*1-4*). Displacement chromatography offers significant potential for simultaneous concentration and purification of biomolecules (*5-12*). In contrast to overloaded elution chromatography, which can often result in significant tailing of the peaks and dilution of the feed, displacement chromatography can produce sharp boundaries and concentrated products during the separation process. Because the process takes advantage of the non-linearity of the adsorption isotherms, a relatively small chromatography column operated in the displacement mode can process higher feed loads, enabling the

0097–6156/93/0529–0027$06.00/0

purification of large amounts of material from dilute biotechnology feeds. While the displacement mode of chromatography has been shown to be a powerful technique for the simultaneous concentration and purification of biomolecules, the elution mode of chromatography predominates industrial applications.

The use of ion-exchange displacement chromatography for the purification of proteins has been investigated by a number of researchers (10,13-23). While these separations resulted in concentration and purification of the bioproducts, the effluent profiles often deviated from the traditional adjacent square wave pattern. Peterson (13) observed "an unexpectedly high concentration of counter-ions" which resulted from the adsorption of the displacer, carboxymethyldextran, during the separation of serum proteins. We have also observed induced salt gradients in a variety of cation and anion exchange displacement separations (20-23). The induced salt gradient, caused by the adsorption of the displacer front, can produce dramatic changes in the behavior of ion-exchange displacement systems.

Clearly, in order to accurately model ion-exchange protein displacement systems, these salt effects must be taken into account. Regnier and co-workers have used a stoichiometric displacement model (SDM) to describe the chromatographic behavior of biopolymers in single component ion-exchange systems (3,24,25). Cysewski et al. (26) have extended this work to non-linear single component ion-exchange systems. To date, multicomponent langmuir adsorption isotherms have been almost exclusively employed to describe non-linear competitive adsorptive systems (27-41). Unfortunately, this formalism is thermodynamically consistent only when the saturation capacities of the proteins are identical (42). Furthermore, salt effects cannot be explicitly accounted for in the Langmuir formalism.

Velayudhan and Horváth (43,44) have used a classical mass-action formalism for the characterization of multicomponent protein binding in ion-exchange systems. Furthermore, they have shown that *multipoint* binding of proteins in ion-exchange chromatography can be accurately represented using this formalism. Brooks and Cramer (45) have presented a refinement of this theory which explicitly accounts for the steric hindrance of salt counter-ions upon protein binding. The steric mass-action ion-exchange formalism, enables the characterization of protein adsorption isotherms unaccounted for in the previous theory. In this chapter, the theory has been used for the prediction of ideal displacement profiles and induced salt gradients. Theoretical predictions are compared to experimental results for a variety of ion-exchange displacement systems.

Theoretical Development

Steric Mass-Action (SMA) Ion-Exchange Equilibrium. In this chapter we present an abbreviated treatment of the SMA formalism. For the complete theoretical development the reader is referred to reference 45. Consider the adsorption of a protein on an ion-exchange resin with a total capacity, Λ, equilibrated with a carrier solution containing salt counter-ions. Upon binding, the protein interacts with v_i stationary phase sites, displacing an equal number of mono-valent salt counter-ions. As seen in Figure 1, the adsorption of the protein also results in the steric hindrance of σ_i salt counter-ions. The stoichiometric exchange of the protein and *exchangeable* salt counter-ions (neglecting the effects of the co-ion) can be represented by,

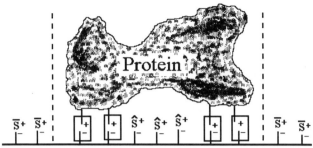

Figure 1. Schematic of protein multi-point attachment and steric hindrance in the SMA formalism. (Reproduced with permission from ref. 44. Copyright 1988 Elsevier Science Publishers BV.)

$$C_i + \nu_i \bar{Q}_1 \Leftrightarrow Q_i + \nu_i C_1 \tag{1}$$

where ν_i is the characteristic charge of the protein. C and Q are the mobile phase and stationary phase concentrations, while the subscripts i and 1 refer to the protein and salt, respectively. The overbar, $\bar{}$, denotes bound salt counter-ions *available* for exchange with the protein. For the sake of simplicity, the equilibrium expressions will be developed for a mono-valent salt counter-ion. In the case of an n-valent salt, ν_i, would represent the ratio of the protein and counter-ion characteristic charges. The equilibrium constant, K_{1i}, for the ion-exchange process is defined as,

$$K_{1i} \equiv \left(\frac{Q_i}{C_i}\right)\left(\frac{C_1}{\bar{Q}_1}\right)^{\nu_i} \tag{2}$$

The total amount of sterically hindered salt ions, $\hat{Q}_1$, *unavailable* for exchange with the protein is given by,

$$\hat{Q}_1 = \sigma_i Q_i \tag{3}$$

where σ_i is the steric factor of the protein. After adsorption of the protein, the total concentration of salt on the stationary phase, Q_1, is given by the following expression:

$$Q_1 = \bar{Q}_1 + \sigma_i Q_i \tag{4}$$

Electroneutrality on the stationary phase requires,

$$\Lambda \equiv \bar{Q}_1 + (\nu_i + \sigma_i) Q_i \tag{5}$$

Substituting equation 5 into equation 2 and rearranging yields the following implicit isotherm:

$$C_i = \left(\frac{Q_i}{K_{1i}}\right)\left(\frac{C_1}{\Lambda - (\nu_i + \sigma_i) Q_i}\right)^{\nu_i} \tag{6}$$

Given the mobile phase concentrations and the bed capacity, equation 6 implicitly defines Q_i, the equilibrium stationary phase concentration of the protein. Once Q_i is determined, the equilibrium stationary phase concentration of the salt, Q_1, is readily calculated from equations 4 and 5. We will refer to this equilibrium between the protein, salt counter-ion and the adsorption bed, as a *single* component isotherm.

The equilibrium formalism described above assumes that the characteristic charge and the steric factor of a given solute are constant and independent of the concentrations of the salt counter-ion and the solute itself. Non-ideal effects such as aggregation or changes in the tertiary structure of the protein are also not accounted for. In addition, the equilibrium constant defined by equation 2 assumes the solution and adsorbed phases act as *ideal* thermodynamic phases allowing the use of concentrations instead of activities (*46*).

Consider the following limiting case. Under overloaded conditions (*i.e.*, $C_i \rightarrow \infty$ and $\bar{Q}_1 \rightarrow 0$) the isotherm (equation 5) reduces to,

$$\lim_{\bar{Q}_1 \rightarrow 0} \quad Q_i = \frac{\Lambda}{\sigma_i + \nu_i} = Q_i^{max} \tag{7}$$

While the stationary phase saturation capacity of solute i, Q_i^{max}, is independent of the mobile phase salt concentration; only at sufficiently low mobile phase salt concentrations does the isotherm in fact approach this limiting value.

From equation 7 we see that the sum of the characteristic charge and the steric factor provides a direct relationship between the bed capacity and the saturation capacity of a given protein. An important feature of the SMA formalism, which distinguishes it from previous theoretical treatments, is the incorporation of the steric factor. Without this additional degree of freedom, the saturation capacity of a given solute is *preordained* by the ratio of the bed capacity and the solute's characteristic charge. Experimentally, this is seldom the case for large biomolecules (*20-22,26*).

Multicomponent Equilibrium. Competitive multicomponent ion-exchange can be formulated as a simple extension of the above single component steric mass-action equilibria. The convention of numbering the species in the sequence of increasing affinity will be adopted here. For a system of n proteins and one salt, we can write n equilibrium expressions representing the stoichiometric exchange of each individual protein with the salt (*i.e.* component 1):

$$C_i + \nu_i \bar{Q}_1 \Leftrightarrow Q_i + \nu_i C_1 \qquad\qquad i = 2,3,...,n+1 \tag{8}$$

The equilibrium constants are define as,

$$K_{1i} \equiv \left(\frac{Q_i}{C_i}\right)\left(\frac{C_1}{\bar{Q}_1}\right)^{\nu_i} \qquad\qquad i = 2,3,...,n+1 \tag{9}$$

The electroneutrality condition on the stationary phase for $n+1$ components becomes,

$$\Lambda \equiv \bar{Q}_1 + \sum_{i=2}^{n+1} (v_i + \sigma_i) Q_i \qquad (10)$$

Equations 9 and 10 provide $n+1$ equations which implicitly define the multicomponent equilibrium for n proteins and the salt counter-ion. In general, the stationary phase concentration of component j can be written as (45),

$$Q_j = \frac{\Lambda \, \alpha_{j1} \, C_j}{\sum_{i=1}^{n+1} (\sigma_i + v_i) \, \alpha_{i1} \, C_i} \qquad j = 1,2,..., n+1 \quad (11)$$

where the separation factor, α_{i1}, is defined as,

$$\alpha_{i1} \equiv \frac{Q_i/C_i}{\bar{Q}_1/C_1} = K_{1i} \left(\frac{\bar{Q}_1}{C_1} \right)^{v_i - 1} \qquad i = 2,3,..., n+1 \quad (12)$$

If Equation 11 is written in terms of mole fractions, the following variable coefficient multicomponent *Langmuir* type isotherm results:

$$y_j = \frac{\alpha_{j1} \, x_j}{1 + \sum_{i=2}^{n+1} [(\sigma_i + v_i) \, \alpha_{i1} - 1] \, x_i} \qquad j = 1,2,..., n+1 \quad (13)$$

where the mole fractions are defined as,

$$x_j = \frac{C_j}{C_{Total}} \qquad y_j = \frac{Q_j}{\Lambda} \qquad j = 1,2,..., n+1 \quad (14)$$

For mono-valent ion-exchange systems without steric effects, Equation 13 reduces to the following stoichiometric exchange isotherm with constant separation factors derived by Helfferich and Klein (27):

$$y_j = \frac{\alpha_{j1} \, x_j}{1 + \sum_{i=2}^{n+1} [\alpha_{i1} - 1] \, x_i} \qquad j = 1,2,..., n+1 \quad (15)$$

Determination of Equilibrium Parameters

Moderately Retained Proteins: Log k' vs. Log Salt Method. The isotherms defined above require the determination of three independent parameters: the characteristic charge of the solute, v_i, the equilibrium constant, K_{1i}, and the steric factor, σ_i. Under linear conditions, well established relationships for protein ion-exchange can be employed to determine the equilibrium constant and the characteristic charge (47-50). The logarithm of the capacity factor (k') can be represented as (44),

$$log\ k' = log\ (\beta\ K_{1i}\ \Lambda^{v_i}) - v_i\ log\ C_1 \tag{16}$$

where the β is the column phase ratio. A plot of the logarithm of the capacity factor versus the logarithm of the mobile phase salt concentration should yield a straight line with the following properties:

$$\text{slope} = - v_i \tag{17a}$$
$$\text{intercept} = log\ (\beta\ K_{1i}\ \Lambda^{v_i}) \tag{17b}$$

Thus, from the linear elution behavior of the solute under various mobile phase salt conditions one is able to determine the characteristic charge, v_i, and the equilibrium constant, K_{1i}. The steric factor, which requires information from the non-linear portion of the isotherm, is determined independently from a frontal experiment. The value of the steric factor, σ_i, can be calculated directly from the breakthrough volume, V_B, using the following expression in concert with the values of the characteristic charge and the equilibrium constant determined from the log k' analysis:

$$\sigma_i = \frac{\beta}{\Pi\ Cf_i}\left(\Lambda - C_1\left(\frac{\Pi}{\beta\ K_{1i}}\right)^{1/v_i}\right) - v_i \tag{18}$$

where, V_o, is the column dead volume and the parameter Π is defined as,

$$\Pi = \left(\frac{V_B}{V_o} - 1\right) \tag{19}$$

Strongly Retained Displacers: Integration of Induced Salt Gradient Method. While the determination of the characteristic charge and equilibrium constant from linear elution data works well for proteins which are moderately retained, solutes which are very strongly adsorbed are difficult to characterize in this fashion. Experimentally, these solutes exhibit regions with very strong retention or no retention at all. To circumvent this experimental difficulty, the characteristic charge can be determined by measuring the change in the salt concentration caused by the frontal adsorption of a solute at a concentration Cf_i. The characteristic charge of the solute is given by,

$$v_i = \frac{\Delta(C_1)_i}{Cf_i} \tag{20}$$

where $\Delta(C_1)_i$ is the increase in the salt concentration caused by the front of solute i.

For solutes which are very strongly adsorbed, the concentration of solute adsorbed during the frontal procedure approaches its saturation capacity. If the mobile phase salt concentration is sufficiently low, the concentration of the adsorbed solute, Q_i, can be assumed to be equal to Q_i^{max}, the saturation capacity of the solute. The steric factor can then be calculated directly from equation 7.

One should be careful to only extract information about the steric factor from a frontal analysis on solutes which are retained enough to approach *complete saturation* of the stationary phase. Typically, the isotherms of highly charged polymers used as displacers in our laboratory approach *square* isotherms that are essentially invariant under low mobile phase salt conditions. Under these conditions the assumption of complete saturation is reasonable and the frontal analysis is justified.

Having determined the characteristic charge and steric factor, a second frontal experiment can then be performed at elevated salt conditions where the stationary phase concentration deviates from its maximum value. The equilibrium constant, K_{1i}, can now be determined from the breakthrough volume of the second front, V_{B2}, using the following expression:

$$K_{1i} = \frac{1}{\beta}\left(\frac{V_{B2}}{V_o} - 1\right)\left(\frac{C_1}{\Lambda - (v_i + \sigma_i)\frac{Cf_i}{\beta}\left(\frac{V_{B2}}{V_o} - 1\right)}\right)^{v_i} \tag{21}$$

Thus, for strongly retained solutes (*e.g.*, displacers) the determination of equilibrium parameters requires data from two frontal experiments. The characteristic charge and steric factor are first determined form a frontal experiment performed under very low salt conditions where the solute *completely* saturates the stationary phase. The equilibrium constant is then determined from a second experiment performed at elevated salt conditions where the stationary phase concentration deviates from its maximum value.

Calculation of Ideal Displacement Profiles

Given a column of sufficient length, displacement chromatography results in the formation of a displacement train containing the feed components in adjacent pure zones. Under such *isotachic* conditions, the species velocities of all the pure components in the displacement train are identical and equal to the velocity of the displacer. In ion-exchange displacement chromatography, the adsorption of the displacer causes an induced salt gradient which travels ahead of the displacer. This increase in salt concentration is *seen* by the feed components traveling in front of the displacer, and results in a depression of the feed component isotherms. The determination of *ideal* displacement profiles reduces to a problem of determining the local salt microenvironment each feed component *sees* in the isotachic displacement train.

The velocity of the displacer, u_d, can be calculated, *apriori*, from the following expression:

$$u_d = \frac{u_o}{1 + \beta\frac{Q_d^*}{Cf_d^*}} \tag{22}$$

where u_o is the chromatographic velocity and the superscript, *, refers to equilibrium values in the isotachic displacement train. The ratio Q_d^*/Cf_d^* is determined from the displacer's *single* component isotherm (equation 6):

$$\frac{Q_d^*}{Cf_d^*} = K_{1d}\left(\frac{\Lambda - (v_d + \sigma_d)\,Q_d^*}{(C_1)_d^*}\right)^{v_d} \tag{23}$$

$(C_1)_d^*$ is the concentration of salt the displacer *sees* (*i.e.*, the carrier salt concentration). Under isotachic conditions, equality of all species velocities implies,

$$\frac{Q_2^*}{C_2^*} = \frac{Q_3^*}{C_3^*} = \ldots = \frac{Q_d^*}{Cf_d^*} \equiv \Delta \tag{24}$$

where Δ is the slope of the displacer operating line.

In a fully developed *ideal* displacement train the feed components are present in adjacent zones of pure material. Thus, the *single* component equilibrium expression (equation 6) can also be used for each feed component in the *ideal* displacement train. Combining equations 6 and 24 yields the following expression:

$$\Delta = \frac{Q_i^*}{C_i^*} = K_{1i}\left(\frac{\Lambda - (v_i + \sigma_i)\,Q_i^*}{(C_1)_i^*}\right)^{v_i} \tag{25}$$

where $(C_1)_i^*$ is the local salt microenvironment *seen* by feed component i in the final isotachic displacement profile. It can be shown (*45*) that this salt concentration is given by the following expression:

$$(C_1)_i^* = \frac{\Lambda - [C_1^\circ + \Delta(C_1)_d]\,\Delta\left(1 + \frac{\sigma_i}{v_i}\right)}{\left\{\left(\frac{\Delta}{K_{1i}}\right)^{1/v_i} - \Delta\left(1 + \frac{\sigma_i}{v_i}\right)\right\}} \tag{26}$$

C_1° is the initial carrier salt concentration. $\Delta(C_1)_d$ represents the increase in the salt concentration due to the introduction of the displacer and is given by $v_d\,Cf_d$. The concentration of feed component i, C_i^*, in the isotachic displacement zone is given by,

$$C_i^* = \frac{\Lambda - (C_1)_i^*\left(\frac{\Delta}{K_{1i}}\right)^{1/v_i}}{(v_i + \sigma_i)\,\Delta} \tag{27}$$

Finally, the volume of the isotachic displacement zone, V_i^*, is given by,

$$V_i^* = Vf\frac{Cf_i}{C_i^*} \tag{28}$$

where Vf is the feed volume. Thus, once the slope of the displacer operating line is determined (*i.e.*, Δ), the local salt microenvironment for each feed component can be determined from equation 26. The concentrations and widths of the ideal isotachic displacement zones are given by equations 27 and 28, respectively.

Experimental

Materials. A Protein-Pak SP-8HR strong cation-exchange column (8 μm, 100 x 5 mm I.D.) and a Protein-Pak QP-8HR strong anion-exchange column (8 μm, 100 x 5 mm I.D.) were donated by Waters Chromatography Division of Millipore (Milford, MA). Sodium chloride, sodium nitrate, sodium phosphate (mono and dibasic) as well as the proteins: α-chymotrypsinogen A, β-lactoglobulin A, β-lactoglobulin B, crude β-lactoglobulin A-B mixture, cytochrome C and lysozyme were purchased from Sigma Chemical Company (St. Louis, MO). The displacers, 50kd dextran-sulfate and 40kd DEAE-dextran were donated by Pharmacia-LKB Biotechnology Inc., (Uppsala, Sweden); protamine sulfate was purchased from ICN Biochemicals (Cleveland, OH).

Apparatus. The chromatographic system employed in this work consisted of a Model LC 2150 HPLC pump (Pharmacia LKB) connected to the columns via a Model C10W 10-port valve (Valco, Houston TX). The column effluent was monitored at 280 nm by a Model 757 spectroflow UV-Vis detector (Applied Biosystems, Ramsey, NJ) and a Model CR601 integrator (Shimadzu, Columbia, MD). Fractions of the column effluent were collected with a Model 2212 Helirac fraction collector (Pharmacia-LKB).

Procedures. A schematic of the chromatograph system employed in the work is illustrated elsewhere (*8*). In all displacement experiments, the columns were sequentially perfused with carrier, feed, displacer and regenerant solutions. Fractions from the column effluent were collected throughout the displacement run and subsequently analyzed. Analytical chromatography was used for the quantification of proteins and displacers. The concentrations of sodium and chloride counter-ions were determined by atomic adsorption and titration, respectively (*20*).

Results and Discussion

Our laboratory has experimentally determined the SMA equilibrium parameters for a variety of proteins and displacer compounds (*20-22*). In addition, we have developed a number of ion-exchange displacement separations using an assortment of displacers (*20-23*). The SMA equilibrium formalism was employed to simulate the displacement purification of α-chymotrypsinogen A and cytochrome C using a 40 kd DEAE-dextran displacer. As seen in Figure 2, the model predicts that perfusion with the DEAE-dextran displacer will result in displacement of the proteins along with an induced salt gradient. Thus, the proteins *see* different salt concentrations in the fully developed displacement zones. The salt concentrations in the α-chymotrypsinogen A and cytochrome C zones are 120 and 116 mM, respectively, in contrast to the 75 mM salt present in the carrier. The SMA model can also be used to generate the adsorption isotherms (Figure 3) of the proteins and displacer under the initial carrier salt conditions, as well as those present in the final isotachic displacement train. As seen in the figure, the isotherm of DEAE-dextran crosses the protein isotherms. Although the DEAE-dextran isotherm does not lie above the protein isotherms, the SMA formalism predicts that this displacement will result in complete separation of the feed proteins.

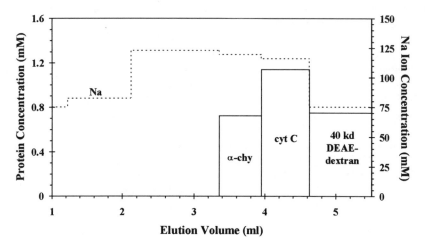

Figure 2. Simulated effluent profile of 40kd DEAE-dextran displacement. (Reproduced with permission from ref. 21)

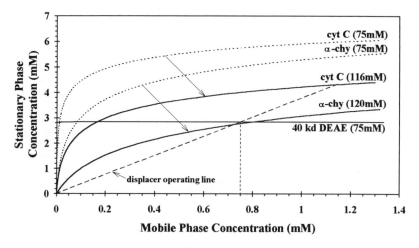

Figure 3. Adsorption isotherms and displacer operating line for 40kd DEAE-dextran displacement. (Reproduced with permission from ref. 21)

The displacement experiment was then carried out to test the SMA model. As seen in Figure 4, the experimental results are in good agreement with the predictions of the model. Both the induced salt gradient and the displacement profiles match well with the theoretical predictions. In addition, an impurity in α-chymotrypsinogen A, not detectable by analytical HPLC in the dilute feed sample, was concentrated and purified in the first fraction of the displacement zone.

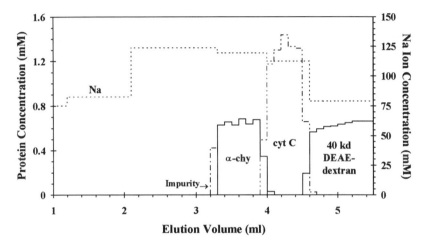

Figure 4.   Effluent profile of 40 kd DEAE-dextran displacement.   (Reproduced with permission from ref. 21)

This displacement differs from the traditional vision of displacement in several important ways: the isotherm of the displacer does not lie above the feed component isotherms; the concentration of displaced protein exceeds the concentration of the displacer; and the induced salt gradient produces different salt microenvironments for each displaced protein. These results demonstrate the ability of the SMA formalism to predict complex behavior present in ion-exchange protein displacement systems.

The displacement purification of β-lactoglobulin A and B has been reported by several investigators (*15,18*). We have used this model separation to examine the efficacy of dextran-sulfate displacers of various molecular weights (*20*). The displacement purification of β-lactoglobulin A and B using a 50 kd dextran-sulfate displacer and the corresponding SMA simulation are shown in Figures 5 and 6, respectively. The SMA model predicts that the dextran-sulfate displacer will result in displacement of the proteins along with an induced salt gradient. Again, the induced salt gradient produced by the displacer results in a depression of the feed component isotherms and the proteins *see* different salt concentrations in the fully developed isotachic displacement zones.

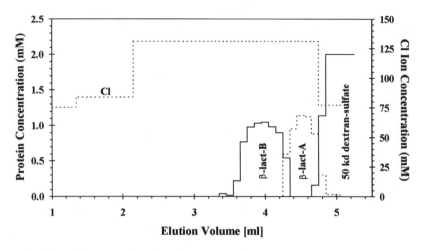

Figure 5. Effluent profile of 50 kd dextran-sulfate displacement. (Reproduced with permission from ref. 21)

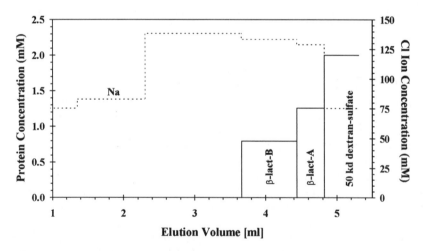

Figure 6. Simulated profile of 50 kd dextran-sulfate displacement. (Reproduced with permission from ref. 21)

As seen in the figures, the experimental results are in qualitative agreement with the predictions of the model. Since β-lactoglobulins A and B are known to aggregate (*51*), the current form of the SMA model is not able to completely describe the adsorption isotherms of these proteins (*20*). Nevertheless, the model still provides a reasonable prediction of the displacement profile under induced salt gradient conditions.

As stated above, our laboratory is in the process of developing a number of displacer compounds for a variety of adsorbent systems. Figure 7 presents the displacement purification of a three component feed mixture composed of α-chymotrypsinogen A, cytochrome C and lysozyme using protamine as the displacer (*22*). As seen in the figure, this relatively inexpensive small protein (5 kd) is an efficient displacer for cation-exchange systems.

The SMA simulation of this separation is shown in Figure 8. As seen in the figure, the experiment is well predicted by the model. The induced salt gradient caused by the displacer results in local salt microenvironments of 74, 72 and 68 mM for the feed components α-chymotrypsinogen A, cytochrome C and lysozyme, respectively. This is in comparison to the initial carrier salt conditions of 25 mM. Clearly, predicting the isotachic displacement profile of this separation using the initial carrier conditions would produce erroneous results. In fact, we have shown that the induced salt gradient caused by the adsorption of the displacer can result in the separation being transformed from the displacement regime to the elution regime (*22*).

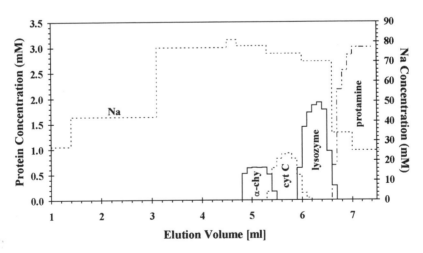

Figure 7. Effluent profile of protamine displacement. (Reproduced from ref. 22. Copyright 1992 American Chemical Society.)

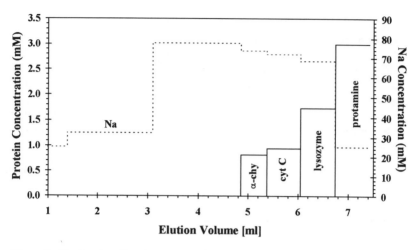

Figure 8. Simulated profile of protamine displacement. (Reproduced with permission from ref. 22. Copyright 1992 American Chemical Society.)

Conclusions

In order to accurately describe the non-linear adsorption behavior of proteins in ion-exchange systems, steric effects as well as salt effects must be taken into account. The theoretical and experimental results presented in this chapter demonstrate the power of the SMA formalism to predict protein adsorption and complex behavior in ion-exchange protein displacement separations. Furthermore, the development of a rapid method for obtaining ideal isotachic displacement profiles with this formalism facilitates methods development and optimization of ion-exchange protein displacement separations.

Notation

C_i	mobile phase concentration solute i (per volume mobile phase)	[mM]
Cf_i	feed concentration solute i	[mM]
C_i^*	isotachic mobile phase concentration solute i	[mM]
C_1^o	carrier salt concentration	[mM]
$(C_1)_i^*$	salt concentration in an isotachic zone containing solute i	[mM]
$\Delta(C_1)_i$	change in salt concentration due to a front of solute i	[mM]
k'	capacity factor	[dimensionless]
K_{1i}	equilibrium constant	[dimensionless]
Q_i	stationary phase concentration solute i (per volume stationary phase)	[mM]
Q_i^*	isotachic stationary phase concentration solute i	[mM]
$\bar{Q}_1$	stationary phase concentration, sterically non-hindered salt	[mM]
$\hat{Q}_1$	stationary phase concentration, sterically hindered salt	[mM]
Q_i^{max}	stationary phase saturation capacity solute i	[mM]

u_d	displacer velocity	[cm/s]
u_o	chromatographic velocity	[cm/s]
V_B	breakthrough volume	[ml]
Vf_i	feed volume solute i	[ml]
V^*_i	volume (width) of the isotachic zone containing solute i	[ml]
V_o	breakthrough volume of an unretained solute	[ml]
x_i	mobile phase mole fraction	[dimensionless]
y_i	stationary phase mole fraction solute i	[dimensionless]

Greek letters

α	separation factor	[dimensionless]
β	column phase ratio $(1-\varepsilon_t)/\varepsilon_t$ (where ε_t is the total porosity)	[dimensionless]
Δ	slope of displacer operating line	[dimensionless]
Λ	stationary phase capacity (mono-valent salt counter-ions)	[mM]
ν	characteristic charge	[dimensionless]
σ	steric factor	[dimensionless]

Acknowledgments

This research was funded by Millipore Corporation and a Presidential Young Investigator Award to S.M. Cramer from the National Science Foundation. The gifts of supplies and equipment from Millipore Corporation (Waters Chromatography Division, Millipore, Milford, MA) and Pharmacia LKB Biotechnology (Uppsala, Sweden) are also gratefully acknowledged.

Literature Cited

1 Regnier, F.E. In *High-Performance Liquid Chromatography of Proteins and Peptides;* Hearn, M.T.W.; Regnier, F.E.; Wehr, C.T., Eds.; Academic Press: New York, NY, 1982.
2 Regnier, F.E. *Science,* 1983, *222*, 245.
3 Kopaciewicz, W.; Rounds, M.A.; Fausnaugh, J.; Regnier, F.E. *J. Chromatogr.,* 1983, *266*, 3.
4 Cramer, S.M.; Subramanian, G. In *New Directions in Adsorption Technology;* Keller, G.; Yang. R., Eds.; Butterworth: Stoneham, MA, 1989, 187.
5 Horváth, Cs.; Nahum, A.; Frenz, J. *J. Chromatogr.,* 1981, *218*, 365.
6 Horváth, Cs.; Frenz, J.; Rassi, Z.E. *J. Chromatogr.,* 1983, *255*, 273.
7 Horváth, Cs. In *The Science of Chromatography;* Bruner, F., Ed.; Journal of Chromatography Library; Elsevier: Amsterdam, 1985, Vol. 32; p179.
8 Cramer, S.M.; Horváth, Cs. *Prep. Chromatogr.,* 1986, *215*, 295.
9 Guichon, G.; Katti, A. *Chromatographia,* 1987, *24*, 165.
10 Subramanian, G.; Phillips, M.W.; Cramer, S.M. *J. Chromatogr.,* 1988, *439*, 341.
11 Phillips, M.W.; Subramanian, G.; Cramer, S.M. *J. Chromatogr.,* 1988, *454*, 1.
12 Cramer, S.M.; Subramanian, G. *Sep. Purif. Methods,* 1990, *19*, 31.
13 Peterson, E. *Anal. Biochem.,* 1978, *90*, 767.
14 Peterson, E.; Torres, A. *Anal. Biochem.,* 1983, *130*, 271.

15 Liao, A.W.; Rassi, Z.E.; LeMaster, D.M.; Horváth, Cs. *Chromatographia,* **1987**, *24*, 881.
16 Subramanian, G.; Cramer, S.M. *Biotechnol. Prog.,* **1989**, *5*, 92.
17 Jen, S.C.; Pinto, N. *J. Chromatogr.,* **1990**, *519*, 87.
18 Jen, S.C.; Pinto N. *J. Chromatogr. Sci.,* **1991**, *29*, 478.
19 Ghose, S.; Matiasson, B. *J. Chromatogr.,* **1991**, *547*, 145.
20 Gadam, S.; Jayaraman, G.; Cramer, S.M. *J. Chromatogr.,* (in press).
21 Jayaraman, G.; Gadam, S.; Cramer, S.M. *J. Chromatogr.,* (in press).
22 Gerstner, J.; Cramer, S.M. *Biotechnol. Prog.,* (in press).
23 Gerstner, J.; Cramer, S.M. (submitted to *BioPharm*).
24 Rounds, M.A.; Reignier, F.E. *J. Chromatogr.,* **1984**, *283*, 37.
25 Drager, R.R.; Reignier, F.E. *J. Chromatogr.,* **1986**, *359*, 147.
26 Cysewski, P.; Jaulmes, A.; Lemque, R.; Sebille, B.; Liladal-Madjar, C.; Jilge, G. *J. Chromatogr.,* **1991**, *548*, 61.
27 Helfferich, F.G.; Klein, G. *Multicomponent Chromatography: Theory of Interference;* Marcel Dekker: New York, NY, 1970.
28 Helfferich, F.; James, D.B. *J. Chromatogr.,* **1970**, *46*, 1.
29 Aris, R.; Amundson, N.R. *Philos. Trans. R. Soc. London Ser. A,* **1970**, *267*, 419.
30 Rhee, H.-K.; Amundson, N.R. *AIChE J.,* **1982**, *28*, 423.
31 Helfferich, F.G. *AIChE Symp. Ser.,* **1984**, *233*, 1.
32 Frenz, J.H.; Horváth, Cs. *AIChE J.,* **1985**, *31*, 400.
33 Helfferich, F.G. *J. Chromatogr.,* **1986**, *373*, 45.
34 Geldart, R.W.; Yu, Q.; Wankat, P.; Wang, N.-H. *Sepn. Sci. & Tech.,* **1986**, *21*, 873
35 Katti, A.M.; Guichon, G.A. *J. Chromatogr.,* **1988**, *449*, 25.
36 Yu, Q.; Wang, N.-H. *Comput. Chem. Eng.,* **1989**, *13*, 915.
37 Golshan-Shirazi, S.; Guichon, G. *Chromatographia,* **1990**, *30*, 613.
38 Khu, J.; Katti, A.; Guiochone, G. *Anal. Chem.,* **1991**, *63*, 2183.
39 Yu, Q.; Do, D.D. *J. Chromatogr.,* **1991**, *538*, 285.
40 Golshan-Shirazi, S.; El Fallah, M.Z.; Guichon, G. *J. Chromatogr.,* **1991**, *541*, 195.
41 Katti, A.M.; Dose, E.V.; Guichon, G. *J. Chromatogr.,* **1991**, *540*, 1.
42 Levan, M.D.; Vermeulen, T. *J. Phys. Chem.,* **1981**, *85*, 3247.
43 Velayudhan, A. *Studies in non-Linear Chromatography;* Thesis; Yale University: New Haven, CT, 1990.
44 Velayudhan, A.; Horváth, Cs. *J. Chromatogr.,* **1988**, *443*, 13.
45 Brooks, C.A.; Cramer, S.M. *AIChE J.,* (in press).
46 Zemaitis Jr., J.F.; Clark, D.M.; Rafal, M.; Scrivner, N.C. *Handbook of Aqueous Electrolite Thermodynamics;* AIChE: New York, NY, 1986.
47 Geng, X.; Regnier, F.E. *J. Chromatogr.,* **1984**, *296*, 15.
48 Velayudhan, A.; Horvath, Cs. *J. Chromatogr.,* **1986**, *367*, 160
49 Regnier, F.E.; Mazsaroff, I. *Biotech. Progress.,* **1987**, *3*, 22.
50 Rahman, A.; Hoffman, N. *J. Chrom. Science,* **1990**, *28*, 157.
51 Whitley, R.D.; Vancott, K.E.; Berninger, J.A.; Wang, N.H.L. *AIChE J.,* **1991**, *37*, 555.

RECEIVED October 13, 1992

Chapter 4

Process Chromatography in Production of Recombinant Products

Walter F. Prouty

Lilly Research Laboratories, Eli Lilly and Company, Lilly Corporate Center, Indianapolis, IN 46285

Process chromatography will continue to play a key role in industrial scale operations involving commercialization of recombinant DNA products. The strategy used in design of a commercial-scale process is dependent on whether the product is a well defined chemical entity, or represents a family of products. For example, human insulin processing involves high resolution techniques because anayltical tools can be used to define insulin as a unique chemical entity. However, protein products with considerable post-translational modifications, especially glycoproteins, are heterogeneous mixtures, just like their naturally occurring counterparts. Processing of these proteins is most efficient when group specific techniques, such as affinity methods, are used. In addition, arguments are given for why process optimization work, especially for a mature process, should focus on the latter steps of the process.

Recombinant technology has made available a diverse array of therapeutic protein products. The quality and safety of these products depend upon a robust and reproducible purification process. Chromatographic methods provide manufacturers with the tools to make highly purified protein products reproducibly from lot to lot. Reproducibility is critical for success in the marketplace and in getting a product and its process approved by regulatory agencies. This article describes some of the advantages of using chromatography in purification. The author also discusses his preferences for resin selection, running conditions, process control, and a strategy for directing process optimization efforts to the latter stages of a process.

Process Design Depends Heavily On Product Source
A therapeutic protein product that is characterized as a single chemical species must be treated differently than one that is a mixture of closely related, but heterogeneous molecules. For example, recombinant proteins produced as intracellular products in prokaryotes, such as E. coli, can be purified to a single chemical species. (That may not be an easy task as several chemical variants are produced both during synthesis by the cell and during manipulation of the protein during purification (1,2). Removal of the initiator amino acid, methionine, or N-formyl methionine, either directly or by cleavage of a leader sequence, has been described elsewhere (3,4)).

0097–6156/93/0529–0043$06.00/0

In contrast, secreted recombinant proteins made in mammalian cells often are subject to post-translational modifications, such as glycosylation, limited proteolysis for processing, or modification of amino acids (5,6). These modifications are not carried out uniformly on all newly synthesized proteins. Overproduction of the recombinant protein can result in synthetic rates that are too high to allow complete processing by the cellular machinery (6). In addition, the carbohydrate chains of glycoproteins display considerable heterogeneity, even though the products appear to have similar or identical biological activities.

The two types of products produced by recombinant technology are also treated differently by regulatory agencies. When the therapeutic agent is a single chemical entity, sensitive analytical tests can result in complete characterization of the product. The heterogeneous secreted product, however, is difficult, at best, to characterize completely. Thus, regulatory authorities argue the "process defines the product" (7). That is, product reproducibility and quality are assured by running a fixed process that has very tight controls on it. [Any change in the process, then, raises the possibility that the product is not exactly the same as that produced before the change.] Such restrictions have important implications in process design as they affect both the choice of chromatographic resin and the vendor of the resin, as will be discussed later.

Protein products that are single chemical entities require the use of high-resolution techniques in the purification process. For example, human insulin (hI) is produced from human proinsulin (hPI) by isolation of the latter from insoluble granule fractions of E. coli (8,9). Proinsulin is cleaved by enzymes to produce insulin and the product is subjected to extensive purification (10). Proinsulin can be purified to a single chemical form and, as such, can be separated from chemical variants such as those containing desamidated asparagine groups (2). Figure 1 shows an HPLC profile of partially purified hPI amplified to illustrate the contaminants in the process stream. At low load (the smaller trace in Figure 1), the product appears to be virtually pure. It is only at much higher loads that the closely eluting and chemically related contaminants are seen. Separation of those closely related species can require high resolution techniques. Chromatography can provide that high resolution step and has been a vital tool in bringing human insulin to the marketplace (2,10,12). It offers an excellent choice of separation techniques that take advantage of subtle differences in proteins based on charge (ion exchange), size, and hydrophobicity (reversed phase and hydrophobic interaction). Since chromatographic purification involves interaction of a part of the protein molecule with the resin surface and usually a limited number of points on the protein are involved in that interaction, chromatography can be exquisitely sensitive to changes altering any of those sites (12). The dilemma for the researcher is that the selectivity of the many different resins on the market is not readily predictable. Choice of a resin to use in a separation must often involve trial and error experimentation with the actual process streams.

The heterogeneity inherent in secreted protein products is illustrated in Figure 2. A wide array of contaminant proteins with varied sizes and isoelectric points are shown in the pre-column sample (Figure 2A). In this experiment, the post-column product purity (Figure 2B) exceeds 90% when assayed with a specific antibody, or by biological activity, even though the stained 2 D gel still shows many spots. All of those spots are product related as seen in Western blot analysis (not shown). Generally, they are the result of heterogeneity of the carbohydrate portion of the glycoprotein, i.e., there is a large reduction in number of spots on deglycosylation (13). Structural heterogeneity of the purified product implies use of high resolution techniques in the purification process is not practical. Instead, group specific methods offer a better choice. Figures 3 and 4 illustrate these two alternatives in using chromatographic methods to purify a heterogeneous biological product: more highly resolving versus group specific procedures. Fractions from ion-exchange chromatography of a secreted product (about 90% pure by bio-assay)

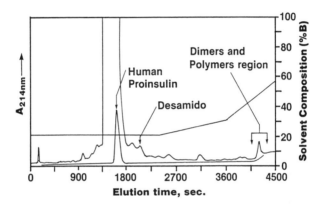

Figure 1. Human Proinsulin is chromatographed on a Zorbax C-8 reversed phase column (*11*). Buffer A was 0.1 M ammonium phosphate, pH 6.8 and Buffer B was acetonitrile. Flow was at 1.0 ml/min and the column run at 45° C. Product was detected by UV absorbance at 214 nm. The smaller tracing was achieved by a ten-fold dilution of the sample into Buffer A.

A   **Pre·Column**

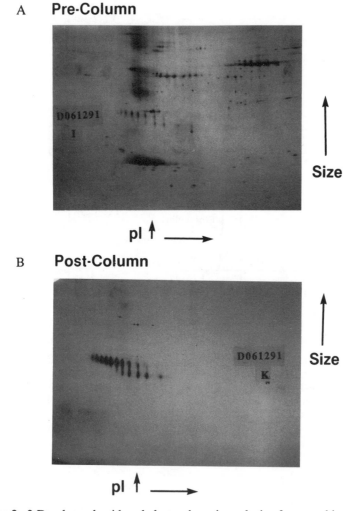

B   **Post·Column**

Figure 2. 2-D polyacrylamide gel electrophoresis analysis of a recombinant product secreted from a mammalian cell (26). The pre-column sample (A) was prepared by filtering the serum-free broth (0.45-lm filter) to effect cell removal and 10X concentration of the mix on a Centricon -10 micro-concentrator (Amicon, Beverley, Ma.). The post column sample (B) was run as is from the column effluent.

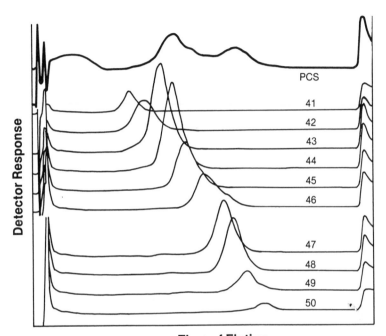

Time of Elution

Figure 3.   Ion-exchange chromatography of a secreted mammalian cell product. A sample of modified tissue plasminogen activator (mtPA) (27) was applied to a cation exchange resin (Mono S, Pharmacia, Upsala, Sweden) and eluted with a shallow gradient of NaCl (0.1 to 0.157M) in 35 mM acetic acid, pH 5, 4.9 M urea and 30% (v/v) Acetonitrile.   Fractions were collected, diluted with low salt buffer, and those containing UV-absorbing material (fractions 41 through 50) were injected onto another Mono S column under conditions identical to the first column. The tracings from the re-injections are shown. The sample identified as "PCS" is the pre-column sample, or starting material. The various fractions elute from the chromatograph as distinct species implying structural or conformational variation.   All fractions from 41 to 50 had enzymatic activity and approximately the same specific activity when measured as amidolytic activity against a small molecule substrate (S2288, Kabi Diagnostica AB, Taljegardsgatan, Sweden) (28).

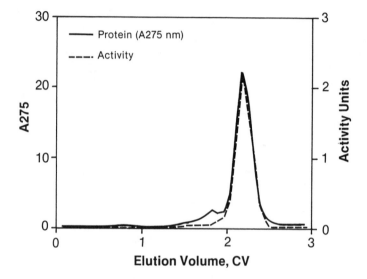

Figure 4. Affinity chromatography of mtPA over a column of Lysine
Sepharose (Pharmacia). The charged column was washed with Tris
(50mM, pH 8.0), sodium chloride (150 mM), Tween 80 (0.01% w/v) buffer
(one column volume), then the product eluted by running a gradient to 500
mM arginine in the tris, salt, detergent wash buffer. Protein was measured
as UV absorbance at 275 nm and activity measured as amidolytic units
using the S2288 substrate as in Figure 3.

showed the preparation could be separated into many discrete products (Figure 3), as shown by the different elution patterns on re-injection on the analytical HPLC. Since these various fractions all contained the same specific activity in a bioassay, such high resolution fractionation results in large yield losses if any of the products is excluded from the mainstream pool. The method is valuable only if separation is desirable. That is, it makes no sense to separate components, if those components are going to be put back together.   A better approach uses affinity techniques, which bind the protein product on the basis of ligand interaction and will be insensitive to many of the chemical variants on the protein that are not associated with the ligand binding site. When the ligand chosen for purification is related to the biological activity of the protein, then a class of proteins with similar biological function will be isolated. Thus, the protein mixture behaves as a single entity, yielding a single peak of activity (Figure 4) when using group specific separation techniques.

Differentiating Early From Late Stages In A Purification Process
Processes for isolation of recombinant proteins can be conveniently broken into an early and a late phase. In the early phase, the goal of the process is twofold: volume reduction, since the product is usually present in the culture broth or cell lysate in low concentration, and clarification, since cell debris is a likely contaminant of product fresh from the fermentor. The early steps of a process are targeted for high volume throughput and product capture (Table 1). High resolution steps are very costly and are not a wise investment at these early processing steps. Late stages of a process may rely on high-resolution methods, if separation of chemically similar species is desired, especially with products of bacterial fermentation, such as human insulin or human growth hormone. The resolution needed, however, depends upon the source of the protein product. As stated above, heterogeneous protein products need not depend upon high-resolution techniques and their extra expense would add undesirable costs to the purification process (Table 1).

In chromatography, high resolution can be achieved by use of small beads (*14,15*). The price for using small beads is that the process must be designed to handle the higher back-pressure generated in the chromatographic columns. The small-bead columns are likely to exceed 15 psi on the production scale columns and thus require special equipment that results in considerable added expense. Above a minimal size and operating pressure (15 psi), buffer tanks must meet American Society of Mechanical Engineers (ASME) codes. Alternatively, the columns can be packed in a short, squat geometry where the column height is much less than the diameter, or in radial flow columns (*16*). However, such columns entail other challenges, such as well-designed sample distribution systems at the inlet and outlet (*17*).

The high costs associated with high resolution may be justified by the potential for increased resolution. In a one-step process, there may be no choice, but usually processes contain multiple steps that take advantage of different properties of the protein such as size, charge, and hydrophobicity. This multiplicity of steps may obviate the need for high resolution techniques or, at least delay their use to the end of the process when clean process streams would result in a longer lifetime in reuse of the column.

Properties Of Chromatographic Resins
Chromatographic resins used in large-scale industrial processes should consist of rigid particles and a uniform size distribution. That is, the resin must neither deform at the pressures generated during the run nor contain very tiny particles that could cause excessive back- pressure (*18*). These properties are important in getting a uniform and well- packed column that can be run at reasonable flow rates. Similarly, the resin should not go through cycles of shrinking and swelling during

column runs or regeneration. Such changes would compromise the integrity of the packed bed and it would be hard to guarantee reproducibility from run to run. Geometric constraints would prevent the shrunken material from swelling back to its original position.

Another important feature of virtually all resins is their ability to participate in secondary interactions. For example, size-exclusion resins can show weak ion-exchange properties (19), as do silica-based reversed-phase resins (20). Different ion-exchange resins from different manufacturers show different selectivity with product streams, even though the functionality may be the same (19). Minimal non-specific binding is often desirable, but on occasion, binding by multiple mechanisms can be helpful if it provides more purification across the step (10,18,21). Again, the choice of the best resin for a product must be based upon lab experiments using the actual process streams.

Economic factors must also be considered in resin choice. Expensive resins can be made economical if manufacturing processes involve multiple uses of the resin in a packed column. Thus, the resin must be hardy enough to withstand rigorous clean-in-place (CIP) procedures, such as acid and base treatment. Base treatment, in particular, solubilizes proteins that may have fouled the resin, and it reduces bioburden (22).

The choice of manufacturer can be as important as the actual type of resin used in a process.   The resin manufacturer must be reliable enough to supply sufficient quantities of resin on a strict time schedule. In addition, the resin must be uniform from batch to batch to provide assurance that the purification process produces a uniform protein product. Finally, resins must not leach monomers or breakdown products into a process stream. A resin manufacturer should assist the resin user by supplying the information how the resin was made, what chemical agents could leach under certain elution conditions, and how to assay for those potential contaminants. To maintain confidentiality, a user could refer to a Drug Master File (DMF) number that the resin manufacturer has submitted to the appropriate regulatory authorities.

The choice of a chromatographic resin determines selectivity, but one can also manipulate the elution mode to optimize further the value of a given resin. The best elution mode depends upon where the step is in the process (Table 2). For secreted products from mammalian cells, resolution is not usually an issue, so adsorption chromatography need involve only on-off mechanisms. Step-elution procedures are easier to design and subject to less equipment breakdown than the more difficult gradient mode of elution. However, manipulation of the gradient shape during elution can provide a more pure product at better yields. A shallow gradient will stretch out the peaks and provide an opportunity to make a mainstream pool that contains more of the desired product as shown in Figure 5.

Displacement chromatography is an alternative elution method (23), but to be very effective the user must carefully define conditions. Since the product in displacement chromatography tends to elute as "square waves", contaminants should not coelute with the product. Removal of the contaminants could result in unacceptably low yields from the step. For example, contaminants coelute with insulin in the displacement run shown in the chromatogram in Figure 6. (Note: the scale used to measure the contaminants is different from that used for the insulin and the contaminant levels are very low compared to the insulin, so the chromatography works quite well.)   The backside of the mainstream pool in Figure 6 would be at fraction 39.   That means some fractions (e. g. fractions #40-42) with high concentrations of insulin would not be included in the mainstream. This is in contrast to the gradient-mode separation shown in Figure 5 where the contaminant elutes during a time of decreasing product concentration. One way to avoid this problem with the displacement method is to continue to search for a better displacing agent. Unfortunately, this must be done by trial and error (24).

Table 1
Resolution Desired in Process Steps

Process Stage	Effect	Resolution	
		Bacterial	Mammalian
Early	Capture/vol. red'n	Low	Low
Late	Purify	High	Group specific

Table 2
Elution Method

Stage	Bacterial	Mammalian	Advantage
Early	Step/grad.	Step/grad.	Step is easy on/off mode-goal of step, e.g. conc?
Late	Gradient	Step/grad.	Desired resol'n?

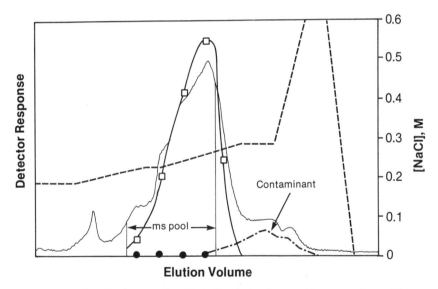

Figure 5. Ion Exchange of an E. coli - derived recombinant product. The protein mixture was charged to a cation-exchange resin (Fast Flow S, Pharmacia, Upsala, Sweden) in 7M urea buffered with 50 mM acetic acid, pH 3.5. The product was eluted with a gradient of NaCl as shown. Protein was monitored by UV absorbance at 280 nm while product and contaminant concentrations were monitored by an analytical HPLC assay. The fractions pooled to make product mainstream (ms pool) are indicated by the vertical lines and represent those fractions with highest specific activity.

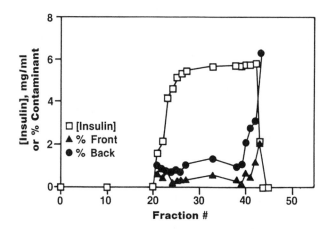

Figure 6.   Displacement chromatography of human insulin (hI) (*18,29*).
Human insulin was charged to a C8 reversed phase resin (Zorbax, DuPont,
Wilmington, De.) at 45.3   g of product per liter of packed resin.   The
starting BHI preparation (97.0% pure by analytical HPLC) was dissolved in
50mM phosphate, pH 2.7 with 2.5% v/v methanol (Buffer A).   The column
was washed with five column volumes of Buffer A and the bound products
were displaced with 10 mM cetyl trimethyl ammonium bromide in Buffer
A. Fractions were collected and analyzed for insulin (expressed as mg/ml)
and percentage contaminant level (expressed as a percent of the insulin
content) by analytical HPLC. The recovery of insulin was quantitative in
this experiment. Note the contaminants elute toward the back of the BHI
peak, but there is some overlap in the points where contaminant
concentration begins to increase and BHI is still at its peak concentration.
The pre-column material (PCS) contained 96.8% BHI with 2.5% frontside
and 0.7% backside contaminants, as measured by HPLC. The mainstream
pool (MS) off of the column contained 98.9% BHI, 0.3% frontside and
0.8% backside contaminants in the analytical HPLC (Reprinted from Ref.
18. Copyright 1991 Marcel Dekker Inc.)

A second way to improve yields is to recycle the sidestream. Recycling poses a potential regulatory problem in that it is more difficult to assure that the recycled product is of exactly the same quality as the mainstream product from the first pass. In addition, the recycle step probably requires a dedicated column for processing. The cost of the recycled product, which includes the extra processing, must be weighed against the cost of merely running more material through from the beginning of the process. The most economical approach is not always intuitively obvious. Often, it is more cost effective, concerning utilities, space, and the time it takes for both step operations and product analysis not to recover sidestreams, but to use those resources to process more single-pass mainstream material from the beginning of the process.

Where To Spend Development Time In Process Improvement
Some of the factors that justify process improvement are: improving product quality, increasing process capacity, and decreasing costs. One of the difficulties of assessing process improvement with respect to product quality is the availability of a suitable assay that can measure the small, incremental effects often seen with the process changes. Use of statistical methods has become a necessity for defining when a process is or is not in control. Statistical methods can also be used to define when variations require action or are just a part of the random variation involved in the evaluation.

Process control charts (25) are used to track the variation in a given process parameter, such as product purity, through a purification step (Figure 7). Variations are introduced at a process step in many ways. For example, mechanical valves have tolerances that mean they may switch at slightly different times after a signal is sent to them, pumps have slight variations and the variations increase with wear, and assays have inherent variability due to both mechanical and procedural manipulations. All steps are also subject, of course, to human error. These human errors involve minor differences in the way steps are carried out. Process control charts allow one to understand the degree of variation introduced by the sum of these individual variations without necessarily understanding the detail of each. In the example shown (Figure 7), the purity of the product from the step is 96.31% (in runs 1 through 80) with the upper and lower control limits (UCL and LCL, respectively) representing three standard deviations of the mean. When an individual point falls outside the UCL or LCL, a nonrandom variation is likely to have occurred in the process and the product of that process step must be suspect. It also warrants immediate investigation as to what caused the variation.

Another use of the chart is to signal when a process is going out of control, i.e., when a trend is developing. Many different criteria are used. A common one is to assume that seven consecutive points on one side of the mean constitute a trend, and warrant careful investigation of the process step.

The chart may also be use to test the effectiveness of a process change. A review of the product purity data subsequent to a process change (Figure 7, at run #80) shows that the process change was favorable both in purity improvement and in reduced variation in the process step. More often, in such process changes, the new values are not changed so dramatically. Then, a large number of runs must be made to assess whether the change resulted in a significant improvement.

When trying to optimize a process on the basis of throughput, or capacity, a convenient method is to chart how much product the process could deliver, given defined step yields for each step in the process (Figure 8). In the process shown, step 7 shows the lowest capacity. Any slipping at that step means that the process would deliver less than the targeted amount of product. Life saving therapeutics, such as human insulin, cannot be run below target, if it means the market cannot be supplied. One approach to improve product throughput is to improve the process at step 7 by either increasing the yield or installing an additional unit to double step

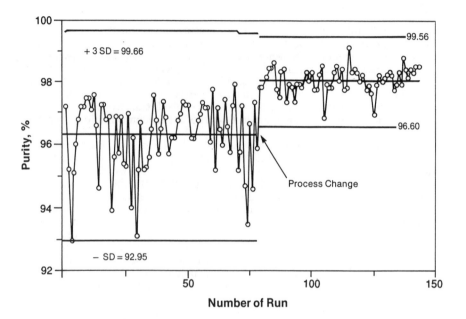

Figure 7. Statistical process control chart for a manufacturing step (30,31).
Purity across the manufacturing step is measured by analytical HPLC
analysis of the product. The upper control limit (UCL) and lower control
limits (LCL) represent three standard deviations from the mean. At run
#80, a process change was made; the shape of the gradient elution from an
ion-exchange chromatography step was changed. A new mean, UCL, and
LCL were calculated for the runs subsequent to the process change to assess
whether the change had a statistically significant impact on product purity.

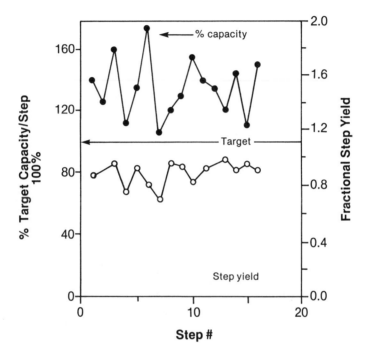

Figure 8. Process capacity at each step of the process. The graph shows the amount of product the process can deliver if all steps are running at the yields indicated. The capacity of a process step is dependent upon the number of operating units at the given step, as well as the size of those units. In chromatography, the throughput can be increased if more material can be put on the column during each cycle (charge capacity), if the elution rate can be increased (increase in flow rate, or change in steepness of elution gradient), or if the size of the columns is increased. Although faster throughput is useful for economic reasons, it can also have a positive effect on product purity, especially if there is any product instability at that step. These factors must be carefully assessed in controlled laboratory studies. Frequently, it is necessary to run nearly full-scale model experiments to study the changes so as to better simulate production conditions.

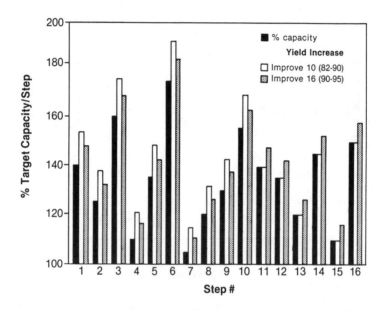

Figure 9. Effect of yield increase at steps 10 or 16 of the process shown in Figure 8. If a yield increase is effected in step 16 (an example is shown for an increase in yield at step 16 from 90 to 95%), then the step is capable of processing 5.6% more material. In addition, the prior step throughput capability is increased by the same factor. In reality, it means less material has to be processed through the previous step to reach the same target goal. Thus, fewer reagents, chemicals, solvents, etc, are needed for all previous steps, if the target is not changed.

capacity. For example, if step 7 is a chromatographic column, a second column could be run, or, alternatively, a larger column installed.

Another way to improve the process throughput, however, is to make process improvements at late steps. This results in a positive effect on all earlier steps. The greater yield of product from step 16 means step 15 and all prior steps need not deliver as much intermediate to get the same throughput. A yield improvement means less intermediate is needed to reach target at all upstream steps (Figure 9), not just the one in which the original improvement was made. Similarly, an improvement late in the process can have a greater impact on costs than earlier process improvements. All previous steps can be run at reduced rates and still maintain the targeted capacity, resulting in less consumption of raw materials, labor, or utilities. The cost of the late stage intermediate product is much higher than earlier intermediates due to the costs of the processing steps and labor to get it there. Thus, it is more cost effective to direct cost saving efforts at increasing yields at the end of the process.

One final factor that is increasingly important in process design is the environmental impact of wastes or noxious chemicals. Improving yields or altering conditions to minimize wastes at the step generating those wastes can provide a solution to the problem. As with throughput, improving yields at the end of the process minimizes the amount of waste stream production from all previous steps. Both approaches are usually taken.

An argument has been made for the importance of tight control and optimization of the latter stages of a process. Regardless of the source of the protein therapeutic, the late process steps are likely to involve chromatography. These chromatographic steps can depend on high resolution, such as when the product is chemically well defined, or on batch methods, such as affinity techniques, when the product is a mixture of closely related products. Large columns (greater than 2000 L) can be run routinely, reproducibly and economically in a manufacturing setting. In addition, it is important to achieve high yields at these steps, both to minimize cost and increase throughput. Finally, there is no method available that can better control quality of a therapeutic protein product while offering a wide choice of selectivity and operating variables than chromatography.

Acknowledgments:
The author thanks his many colleagues in the Biosynthetic Development Department at Eli Lilly for their many fruitful and stimulating discussions and suggestions. In particular, I thank G. Kelly for the Cation Exchange data generated from the secreted mammalian cell product, P. Atkinson and T.Vieira-Brisson for development of the affinity method used in Figure 4, and R. Wilkens for many informative discussions regarding the use of Statistical Process Control (SPC) in control of the Humulin production process.

Literature Cited
(1) Stadtman, E. R. *Biochemistry* **1990**, *29(27)*, 6323.
(2) Prouty, W. F., *BioTech USA, The sixth Annual Industry conference and Exhibition.* BIOTECH Proceedings, San Francisco, Calif. Oct. 2-4, **1989** (Nature Publishing Co., New York, N.Y.) p. 224.
(3) Marston, F.A.O. *Blochem J.* **1986**, *240*, 1.
(4) Nagai, K.; Thogerson, H. C. *Methods Enzymol.* **1987**,*153*, 461.
(5) Hokke, C. H.; Berwerff, A. A.; van Dedem, G.W.K.; van Oostrum, J., Karerling, J.P.; Vliegenthart, J.F.G. *FEBS Lett.* **1990**, *275(1, 2)*, 9.
(6) Yan, S - C.; Grinnell, W.; Wold, F . *Trends in Biochemical Sciences (TIBS)*[**1989**,*14*, 264.
(7) Hammond, P. M.; Atkinson, T.; Sherwood, R. F.; Scawen, M. D. *BioPhann* **1991**, *4(5)*, 30.

(8) Frank, B. H.; Chance, R. E. In*Therapeutic agents produced by genetic engineering* Editors, Joyeaus, A., Leygue, G., Moore, M., Roncacci, R., Schmelk, P.H. "Quo Vadis"? Symposium, Sanofi Group, May 29-30, **1985**. Toulouse-Laberge, France. p. 137.
(9) Frank, B. H.; Chance, R. E. Muench. Med Wochen~chr. 1983, 125, (suppl. 1), 14.
(10) Kroeff, E. P.; Owens, R.A.; Campbell, E. L.; Johnson, R. D.; Marks, H. I. *J. Chromatogr.* **1989**, *461*, 45.
(11) DiMarchi, R. D.; Long, H. B.; Kroeff, E. P.; Chance, R. E. In *High Performance liquid Chromatoeraphy in Biotechnoloegy* Hancock, W. S., Ed.; John Wiley and Sons: New York, **1990**; pp. 181-190.
(12) Rounds, M. A.; Kopaciewicz, W.; Re~er, F.E. *J. Chrornatogr.* **1986**, *362*, 187.
(13) Yan, S-C.B.; Razzano, P.; Chao, B.; Walls, J. D.; Berg, D.T.; McClure, D. B.; Crinnell, B. W. *Bio/Technology* **1990**, *8(7)*, 655.
(14) Johnson, E.; Stevenson, B. Basic *Liquld Chromatography*; Varian Associates: Palo Alto, Calif., **1978**; pp. 15-39.
(15) Rahn, P.; Joyce, W.; Schratter, P. *Am. Biotechnol Lab.* **1986**, *4(7)*, 34.
(16) Saxena, V.; Subramanian, K.; Saxena, S.; Dunn, M. *BioPharm* **1989**, *2(3)*, 46.
(17) Levine, H. L. Presented at the Conference on Frontiers in Bioprocessing, Boulder, Co., June **1987**.
(18) Prouty, W. F. *Production-Scale Purification Processes in Drug Biotechnology Regulation: Scientific Basis and Practtces*: Chiu, Y-Y.H.; Gueriguian, J. L., Eds.; MIarceI Dekker: New York, **1991**; pp. 221-62.
(19) Regnier, R. E. *J. Chromatogr.* **1987**, *418*, 115.
(20) Barford, R. A. In *High Performance Liquid Chrornatoeraphy fn Biotechnology;* Hancock, W. S., Ed.; John Wiley and Sons: New York, **1990**; pp. 63-77.
(21) Horvath, C.; El Rassi, Z. *Chromatogr. Forum* **1986**, *1(3)*, 49.
(22) Cleaning in Place in Perspective; *Downstream,* vol. 9 (a publication of Pharmacia), **1990**; pp. 16-20.
(23) Su6ramanian, G.; Phillips, M. W.; Cramer, S. M. *J. Chromatogr.* **1988**, *439(2)*, 341.
(24) Ghose, S.; Mattiasson, B. *J. Chromatogr.* **1990**, *547*, 14S.
(25) GopaI, C.; Mlodozeniec, A.; Holrnes, D.S.. S. *Pharm.Technol.* **1991**, *15(1)*, 32.
(26) Anderson L. *Two Dimensional Electrophoresis. Operation of the ISO-DALT System;* Large Scale Biology Press: Washington, D.C., **1988**; vol. *12,* Pp. 570-75.
(27 Burck, P. J.; Berg, D. H.; Warrick, M. W., *J. Biol. Chem.* **1990**, *265*, 5170.
(28) Wallen P.; Pohl, G.; Bergsdorf, N.; Ranby, M.; Ny, T.; Jornvall, H. *Eur. J. Biochem* **1983**, *123(3)*, 681.
(29) Vigh, C.; Varga-Pouchon, Z.; Szepsi, G.; Gazdag M. *J. Chromatogr.* **1987**, *386*, 353.
(30) Amsden, R. T.; Butler, H. E.; Amsden D. M. *SPC Simpllfled: Practical Steps To Quality:;* Kraus International Publications: White Plains, N.Y., **1986**.
(31) Ishikawa, K. *Guide to Quality Control;* Asian Productivity Organization, Quality Resources: White Plains, N.Y., **1990**.

RECEIVED October 9, 1992

Chapter 5

Preparative Reversed-Phase Sample Displacement Chromatography of Peptides

R. S. Hodges, T. W. L. Burke, A. J. Mendonca, and C. T. Mant

Department of Biochemistry and the Medical Research Council of Canada Group in Protein Structure and Function, University of Alberta, Edmonton, Alberta T6G 2H7, Canada

This report describes the early development and current applications of a novel method for highly efficient preparative-scale reversed-phase purification of peptides, termed sample displacement chromatography (SDC). This method is based upon solute-solute displacement by components of a peptide mixture following high sample loading in a 100% aqueous mobile phase, and is characterized by the major separation process taking place in the absence of organic modifier. Highly efficient use of stationary phase capacity, coupled with low flow-rates, enable the preparative purification of high sample loads on analytical columns and instrumentation with high yields of purified product concomitant with low solvent consumption. The potential value of SDC for the pharmaceutical industry was examined by its application to the purification of a five-residue synthetic peptide, representing part of the sequence of luteinizing hormone-releasing hormone. Purification of 100 mg of the peptide on a C_8 stationary phase (300 x 4.6 mm I.D.), carried out by three variations of SDC (a multi-column approach, extended isocratic elution in the absence of organic modifier and isocratic elution in the presence of low levels of organic modifier) produced up to 97.3% yields of purified product in a short run time and with minimal solvent consumption. The simplicity of SDC, coupled with high yields of purified product, minimization of the use of potentially toxic reagents and solvents and the potential for straightforward scale-up, points to a promising future for this technique.

The growing use of synthetic peptides in biochemistry, immunology and in the pharmaceutical and biotechnology industries has stimulated a concomitant increase in requirements for rapid and efficient preparative purification methods. Considering that the impurities encountered during peptide synthesis (deletion, terminated or chemically modified peptides) are usually closely related structurally to the peptide of interest, and, hence, often pose difficult purification problems, most preparative separations of synthetic peptides take advantage of the excellent resolving power and volatile mobile phases of reversed-phase chromatography (RPC) in gradient elution mode (1,2). However, the

0097–6156/93/0529–0059$06.00/0

gradient elution mode of RPC is handicapped by relatively poor utilization of the stationary and mobile phases (3,4). Thus, large-scale gradient elution separations, typically involving aqueous trifluoroacetic acid (TFA) to TFA-acetonitrile gradients (1,2) of peptides closely related structurally generally require the employment of larger column volumes in order to maintain satisfactory levels of sample load and product yield. This, in turn, leads to higher operating costs per unit of purified product in terms of packings, equipment and solvent consumption.

A perceived need for simpler, more efficient and reliable methods for preparative separations of peptides led this laboratory to consider ways of modifying various parameters to overcome the many disadvantages frequently encountered in traditional preparative methods: (1) *Column size:* a more efficient utilization of stationary phase capacity would allow the employment of smaller column volumes, while maintaining high sample load; (2) *Solvent consumption:* consumption of expensive HPLC solvents (water, organic modifier) could be cut back considerably if mobile phase flow-rates could be reduced; (3) *Organic modifier:* if the role of the organic modifier in the separation process could be reduced, this would also reduce the consumption of large volumes of frequently toxic and expensive solvents; (4) *Fraction analysis:* reduction of the number of fraction analyses required to locate the product(s) of interest following preparative purification would save valuable time and solvents; (5) *Product concentration:* the large column volumes and flow-rates typical of traditional preparative gradient-elution methods frequently result in dilution of products in large fraction volumes. Thus, any way of concentrating the same level of product in significantly smaller fraction volumes would greatly simplify subsequent product work-up; (6) *Instrumentation:* most researchers would probably desire to carry out laboratory-scale preparative separations, if possible, on existing analytical equipment.

We report here the development of a novel method for highly efficient preparative-scale reversed-phase purification of peptides on analytical columns, termed sample displacement chromatography (SDC). This approach is characterized by the major separation process taking place in the absence of organic modifier. Following a review of the early development of SDC (3-7), we present current applications of the method to the preparative purification of a pharmaceutically important synthetic peptide, underlying once again the potential of SDC for the pharmaceutical industry.

Principles of Sample Displacement Chromatography (SDC)

Since peptides favour an adsorption-desorption method of interaction with an hydrophobic stationary phase (1,2), under normal analytical load conditions an organic modifier is typically required for their elution from the reversed-phase column. However, when such a column is subjected to high loading of a peptide sample mixture dissolved in a 100% aqueous mobile phase, there is competition by the sample components for the adsorption sites on the reversed-phase sorbent, resulting in solute-solute displacement. The more hydrophobic peptide components compete more successfully for these sites than less hydrophobic components, which are displaced and quickly eluted from the column. Thus, operation in sample displacement mode is simply employing the well-established general principles of displacement chromatography (8-10) without using a displacer (3-7).

A crude peptide mixture typically produced by solid-phase peptide synthesis may contain not only the desired product, but also hydrophilic and/or hydrophobic synthetic peptide impurities. Situations where a desired peptide product is the most hydrophilic or most hydrophobic component of a peptide mixture offer the most convenient model system for illustrating the SDC process.

Displacement of Hydrophilic Impurities by Desired Hydrophobic Peptide Component. Figure 1 demonstrates SDC of a five-decapeptide mixture, where peptide 5 is the desired product (P_5) and peptides 1 to 4 (I_1 to I_4, respectively) represent hydrophilic impurities. The hydrophobicity of the peptides increases only slightly between peptide 2 and peptide 5 (sequences shown in Figure 1) - between 2 and 3 there is a change from an α-H to a β-CH$_3$ group, between 3 and 4 there is a change from a β-CH$_3$ group to two methyl groups attached to the β-CH group, and between 4 and 5 there is a change from an α-H to an isopropyl group attached to the α-carbon - enabling a precise determination of the potency of preparative RPC in sample displacement mode. The product, P_5, represented only 33% of the total 21 mg of the five-peptide sample mixture (analytical profile shown in Figure 1A) loaded onto the C$_8$ column (30 x 4.6 mm I.D.). The protocol for SDC of the sample mixture involved displacement in water (0.05% aqueous TFA) of the hydrophilic impurities (I_1-I_4) by the more hydrophobic P_5, which should remain bound to the column. Thus, following isocratic elution with 0.05% aqueous TFA at 1 ml/min, during which the major separation took place and all hydrophilic impurities were eluted from the column (Figure 1B), a gradient wash was initiated after 40 min (1% acetonitrile/min) to remove the product from the column. Following fraction analysis, Peak II (Figure 1B) was found to be pure P_5 (Figure 1D) and accounted for 93% of total P_5 recovered. Peak I, the unretained fraction (Figure 1B), contained the bulk of hydrophilic impurities I_1 to I_4 (Figure 1C).

Displacement of Desired Hydrophilic Peptide Component by Hydrophobic Impurities. Figure 2 demonstrates SDC of a four-decapeptide mixture, where peptide 2 is the desired product (P_2) and peptides 3 to 5 (I_3 to I_5, respectively) represent hydrophobic impurities. The product, P_2, represented 50% of the total 6 mg of the four-peptide sample mixture (analytical profile shown in Figure 2A) loaded onto the C$_8$ column (30 x 4.6 mm I.D.). Following the same SDC protocol as described in Figure 1 (SDC elution profile shown in Figure 2B), Peak I was found to contain pure P_2 (Figure 2C) and accounted for 99% of recovered P_2. The bulk of the impurities, I_3-I_5, remained bound to the column as Peak II (Figure 2D) during elution with 0.05% aqueous TFA, and were only removed by the addition of acetonitrile to the mobile phase.

A short (30 mm in length) column was used to illustrate the basic principles of SDC in order to limit the amount of material required to saturate the hydrophobic stationary phase. The results of Figures 1 and 2 demonstrate that impressive yields of pure peptide products may be obtained even on such small columns.

Development of Multi-Column Approach to SDC of Peptides

The separations demonstrated in Figures 1 and 2, where the desired peptide product is the most hydrophobic or hydrophilic component, respectively, represent the simplest application of SDC. The next step in the development of SDC as a preparative tool required the design of a strategy to deal with the more difficult, and more realistic, separation problem of a desired peptide product contaminated with both hydrophilic and hydrophobic impurities. In addition, scale-up to higher sample loads was a logical, and necessary development. The latter requirement also highlighted the problem of optimizing sample load when applying preparative SDC to peptide separations on single columns. For instance, from Figure 1, if the sample load was too high, product (P_5) may appear in the breakthrough fraction, contaminated with displaced hydrophilic impurities (I_1 to I_4); if the sample load was too low, hydrophilic impurities may remain on the column. From Figure 2, if the sample load was too high, displaced impurities

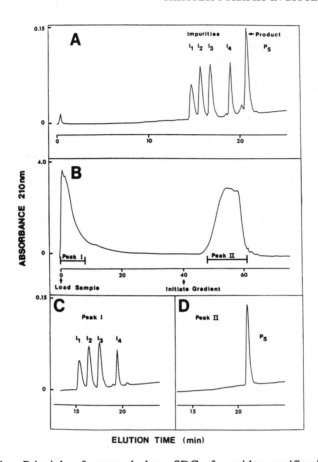

Figure 1.   Principle of reversed-phase SDC of peptides: purification of peptide "product" (P5) from hydrophilic peptide "impurities" (I1, I2, I3, I4). Column: Aquapore RP300 C8, 30 x 4.6 mm I.D., 7-μm particle size, 300-Å pore size (Pierce, Rockford, IL, U.S.A.). HPLC instrument: Varian Vista Series 5000 liquid chromatograph (Varian, Walnut Creek, CA, U.S.A.) coupled to an Hewlett-Packard (Avondale, PA, U.S.A.) HP1040A detection system, HP85B computer, HP9121 disc drive, HP2225A Thinkjet printer and HP7470 plotter.   Panel A: analytical separation profile of peptide mixture; conditions, linear AB gradient (1% B/min) at a flow-rate of 1 ml/min, where eluent A is 0.05% aqueous TFA and eluent B is 0.05% TFA in acetonitrile. Panel B: preparative separation profile of peptide mixture; conditions, isocratic elution with 100% eluent A for 40 min at a flow-rate of 1 ml/min, followed by linear gradient elution at 1% B/min; sample load, 21 mg, consisting of 7.0 g of P5 and 3.5 mg of each of I1, I2, I3 and I4 dissolved in 500 μl of eluent A.   Panels C and D demonstrate analytical elution profiles (see Panel A for conditions) of Peaks I and II (Panel B), respectively.   The subscripts of I1, I2, I3, I4 and P5 denote peptides 1-5, respectively.   The basic sequence of the peptide analogues is Arg-Gly-X-X-Gly-Leu-Gly-Leu-Gly-Lys where X-X is substituted by Gly-Gly (peptide 2), Ala-Gly (peptide 3), Val-Gly (peptide 4) or Val-Val (peptide 5).   The peptides all contain an N^α-acetyl group and a C^α-amide group, except for peptide 1, which is the same as peptide 3 save for a free α-amino group.   Reprinted from reference 3, with permission.

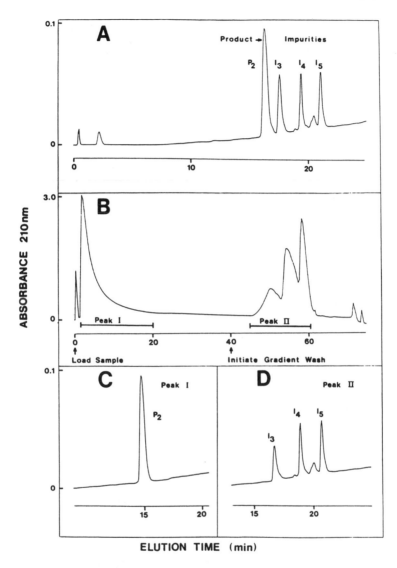

Figure 2.   Principle of reversed-phase SDC of peptides: purification of peptide "product" (P_2) from hydrophobic peptide "impurities" (I_3, I_4, I_5). Column and HPLC instrument: same as Figure 1.   Panel A: analytical separation profile of peptide mixture; conditions, linear AB gradient (1% B/min) at a flow-rate of 1 ml/min, where eluent A is 0.05% aqueous TFA and eluent B is 0.05% TFA in acetonitrile.   Panel B: preparative separation profile of peptide mixture; conditions, isocratic elution with 100% eluent A for 40 min at a flow-rate of 1 ml/min, followed by linear gradient elution at 1% B/min; sample load, 6 mg, consisting of 3 mg of P_2 and 1 mg of each of I_3, I_4 and I_5 dissolved in 100 μl of eluent A.   Panels C and D demonstrate analytical elution profiles (see Panel A for conditions) of Peaks I and II (Panel B), respectively.   The subscripts of P_2, I_3, I_4 and I_5 denote peptides 2-5, respectively.   The sequences of the peptides are shown in Figure 1. Reprinted from reference 3, with permission.

may contaminate the displaced peptide product (P_2) in the breakthrough fraction; if the sample load was too low, some peptide product may remain on the column, contaminated with hydrophobic impurities (I_3 to I_5). The situation would be even more complex with both hydrophilic and hydrophobic peptide impurities present. These concerns prompted the development of a multi-column approach to preparative SDC of peptides (6).

Figure 3 illustrates this multi-column strategy through its application to the purification of 100 mg of a synthetic decapeptide. In addition to the desired peptide product, P, the crude peptide mixture also contained significant levels of both hydrophilic (I_1), and hydrophobic (I_2 to I_5) impurities, as shown in the analytical run (Figure 3, top profile).

The 100 mg sample (in 0.1% aqueous TFA) was applied to a multi-column set-up, consisting of ten 30 x 4.6 mm I.D. C_8 column segments (columns 1-10) in series. Isocratic elution with 0.1% aqueous TFA (at a flow-rate of 0.5 ml/min) was continued up to a total run time of 50 min, which included the time taken for sample loading.

Following SDC and elution of the individual column segments, the distribution of sample components through the ten column segments was determined by analytical runs on the multi-column set-up (Figure 3, bottom profiles). The breakthrough fraction (0) and column 1 (at the multi-column outlet) contained only hydrophilic impurities (I_1). Column 10 (at the multi-column inlet) contained only hydrophobic impurities (I_2 to I_5), while columns 8 and 9 contained a small amount of peptide product contaminated with hydrophobic impurity, I_2.

Columns 2-7 contained pure product, the great majority of which was in columns 3-7. Of the total amount of desired peptide product loaded onto the column, 90% was recovered, with 81% of the peptide product isolated as pure peptide.

The utilization of ten C_8 column segments (instead of just the one segment employed in Figures 1 and 2) enabled the application of the substantial sample load of 100 mg on what is effectively a standard analytical column (300 x 4.6 mm I.D.). From Figure 3, very efficient use of the column capacity had been made during the SDC separation. In addition, only eleven fractions (the breakthrough fraction and the fractions obtained from individual elution of the ten column segments) had to be analyzed, which greatly simplifies the fraction analysis compared to traditional preparative gradient elution chromatography.

One important point to note from Figure 3 is that, although the sample load was not sufficient to displace all hydrophilic impurities from the multi-segmented column (an even higher sample load would have been required to achieve this), pure product is still obtained, since the crude sample components have been distributed along a chromatography bed divided into segments, rather than along a continuous packed bed as was the case for Figures 1 and 2. Thus, even at unoptimized (i.e., too low or, for that matter, too high) sample loads, significant levels of pure product may still be obtained, highlighting the flexibility of this approach.

Application of Preparative Reversed-Phase SDC to the Purification of Pharmaceutically-Important Peptides

In order to investigate the potential value of the sample displacement approach to the pharmaceutical industry, preparative SDC was now applied to the purification of a five-residue synthetic peptide [Glp-His-Trp-Ser-Tyr, denoted LHRH (1-5)], representing part of the sequence of luteinizing hormone-releasing hormone (LHRH). In addition to its pharmaceutical significance, this peptide represented an interesting challenge due to its relatively low solubility in an aqueous mobile

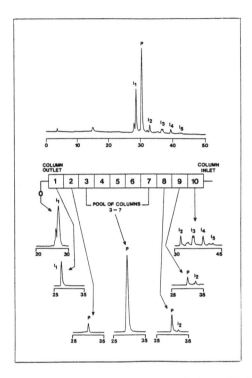

Figure 3. Multi-column approach to reversed-phase SDC of peptides: purification of 100 mg of a synthetic decapeptide crude mixture. Column: ten 30 x 4.6 m I.D. Aquapore RP300 C_8 (7-μm particle size, 300-Å pore size; Pierce, Rockford, IL, U.S.A.) column segments in series. The numbering of the ten columns (or fractions) starts at the column segment closest to the detector. Thus, column 1 (fraction 1) is at the multi-column outlet, while column 10 (fraction 10) is at the inlet. The fraction marked 0 is the breakthrough fraction. HPLC instrument: Varian Vista Series 5000 liquid chromatograph (Varian, Walnut Creek, CA, U.S.A.) coupled to a Hewlett-Packard (Avondale, PA, U.S.A.) HP1040A detection system, HP9000 Series 300 computer, HP9133 disc drive, HP2225A Thinkjet printer and HP7440A plotter. Conditions: (1) *analytical separation* (top elution profile) of the peptide mixture on the multi-column set-up [linear AB gradient (1% B/min and 1 ml/min), where eluent A is 0.1% aqueous TFA and eluent B is 0.1% TFA in acetonitrile]; (2) *column equilibration, sample loading* and the *preparative SDC run* (flow-rate=0.5 ml/min; run time = 50 min) were all carried out by isocratic elution with 0.1% aqueous TFA; (3) following isocratic elution of each individual column segment with 25% (v/v) aqueous acetonitrile containing 0.1% (v/v) TFA, *fraction analysis* of the resulting peptide solutions was carried out by linear AB gradient elution (1% B/min at 1 ml/min) on the multi-column set-up). The detection wavelength was 210 nm for both analytical and preparative runs. The analytical elution profiles at the bottom show the peptide components retained on each individual column segment (columns 1-10) following the SDC run. The peaks are all correctly proportioned. P is the desired product; the other peaks are hydrophilic (I_1) and hydrophobic (I_2-I_5) impurities. The sequence of the peptide product is Ac-Arg-Gly-Val-Val-Gly-Leu-Gly-Leu-Gly-Lys-amide, where Ac denotes N^α-acetyl and amide denotes C^α-amide. Reprinted from reference 6, with permission.

phase based on 0.1% TFA. Also, TFA is not the acidic reagent of choice since the trifluoroacetate salt of the purified peptide (due to ion-pairing of the trifluoroacetate counterion with positively charged residues) is undesirable as a pharmaceutical product. Thus, the mobile phase of choice for preparative reversed-phase SDC of LHRH(1-5) was 5% (v/v) aqueous acetic acid. Acetic acid is occasionally employed as a component of mobile phase systems for RPC of peptides (11), the acetate counterion being more hydrophilic than trifluoroacetate and producing acetate salts of peptides containing positively charged groups. In addition, this particular peptide was more soluble in 5% acetic acid.

Standard Multi-Column Approach to Preparative SDC of LHRH (1-5). Figure 4 illustrates the results of applying the multi-column SDC strategy (on the same ten-column set-up described in Figure 3) to the purification of 100 mg of crude LHRH (1-5). Figure 4A shows the analytical profile of the crude mixture, where the desired peptide product, P, is contaminated by both hydrophilic and hydrophobic impurities (I).

The sample (in 8 ml of 5% aqueous acetic acid) was loaded onto the column at a flow-rate of 0.2 ml/min (i.e., 40 min was required for total loading of the sample), followed by further isocratic elution at the same flow-rate with 5% aqueous acetic acid for 30 min, for a total run time of 70 min. A previous study has demonstrated that low flow-rates enhance overall yield of pure product (6).

Figure 4B shows the distribution of sample components through the ten-column segments following SDC. The breakthrough fraction (BT; inset profiles) contained only hydrophilic impurities. Column 1 (C1; inset profiles) at the multi-column outlet, contained product (10.6% of total product recovered) as well as hydrophilic impurities. Columns 9 and 10 (C9 and C10, the latter being at the multi-column inlet) contained the hydrophobic impurities and a small amount (3.2%) of recovered product. Columns 2 to 8 (C2 to C8) contained pure product only, representing 86.2% of product recovered and underlining a very successful purification of LHRH (1-5).

Peptide Elution in Preparative SDC of LHRH (1-5) in the Absence of Organic Modifier. From Figure 4, the presence of desired product on column 1 (C1) indicated that the run was carried out with close to optimum column loading, i.e., where the product displaces totally all hydrophilic impurities into the breakthrough fraction (BT), while all product remains on the column. In an attempt to avoid the use of organic modifier to elute the product from the individual column segments, the run time was now increased, with a view not only to complete the displacement of hydrophilic impurities from the multi-column set-up, but also to elute the product from the column with the 100% aqueous mobile phase.

Figure 5 illustrates the effect of increased run time on preparative SDC of 100 mg of crude LHRH (1-5) (analytical profile shown in Figure 5A).

The sample was loaded onto the column in 5% aqueous acetic acid as before (Figure 4), followed by further isocratic elution with the same mobile phase for 140 min, for a total run time of 180 min. Unlike the previous run (Figure 4), 2-min (i.e., 0.4 ml) fractions of the column effluent were collected as soon as sample loading commenced.

Figure 5B shows the distribution of sample components in the breakthrough fractions, as well as through the ten-column segments. The major observation here is that fully 88.8% of recovered product has been displaced from the column, with pure LHRH (1-5), representing 83.4% of recovered product, found in fractions 38-90 (in a total volume of only 21.2 ml). Only 11.2% of product remained on the multi-segmented column, 10.9% recovered as pure

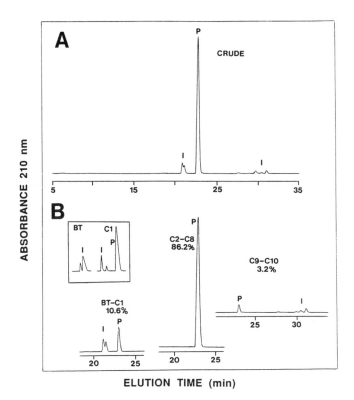

Figure 4. Reversed-phase SDC of 100 mg of crude LHRH (1-5). Column: same as Figure 3. The numbering of the ten columns starts at the column segment closest to the detector. Thus, column 1 (C1) is at the multi-column outlet, while column 10 (C10) is at the inlet. The fraction marked BT is the breakthrough fraction. Sample: 100 mg of crude LHRH (1-5) (sequence, Glp-His-Trp-Ser-Tyr, where Glp denotes pyroglutamic acid), dissolved in 8 ml of 5% (v/v) aqueous acetic acid; the desired peptide product is denoted P and impurities are denoted I. HPLC instrument: same as Figure 3. Conditions: (1) *the analytical elution profile* (panel A) of the crude peptide on the multi-column set-up [linear AB gradient (1% B/min and 1 ml/min), where eluent A is 0.1% aqueous TFA and eluent B is 0.1% TFA in acetonitrile]; (2) *column equilibration* was carried out with 5% (v/v) aqueous acetic acid; (3) following *sample loading* (8 ml sample volume at a flow-rate of 0.2 ml/min = 40 min), the *preparative SDC run* (flow-rate = 0.2 ml/min) was continued by isocratic elution with 5% (v/v) aqueous acetic acid for a total run time of 70 min; (4) following isocratic elution (0.5 ml/min) of each individual column segment (C1-C10) with 1 ml of 25% (v/v) aqueous methanol containing 5% (v/v) acetic acid, *fraction analysis* of the resulting peptide solutions was carried out by linear AB gradient elution (1% B/min at 1 ml/min) on the multi-column set-up described in Panel A. The detection wavelength was 210 nm for both analytical and preparative runs. The analytical elution profiles shown in panel B represent peptide components contained in three sets of pooled fractions, together with the proportion of product, P (expressed as a percentage of total product recovered) found in each pool. The inset profiles show the peptide components individually found in the breakthrough fraction (BT) and retained on column 1 (C1). The peaks in the analytical profiles are all correctly proportioned.

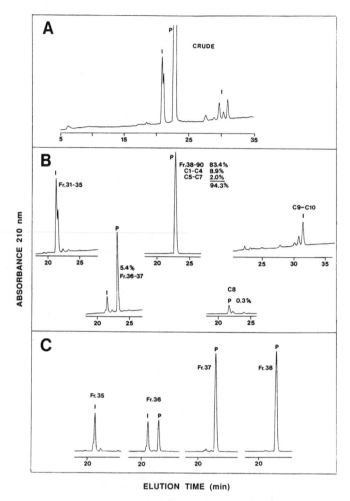

ELUTION TIME (min)

Figure 5. Peptide elution during reversed-phase SDC [100 mg of crude LHRH (1-5)]. Column, HPLC instrument and sample preparation: same as Figure 4. Conditions: (1) the *analytical elution profile* (panel A) of the crude peptide (peptide product, P; impurities, I) was obtained as described in Figure 4; (2) *column equilibration,* same as Figure 4; (3) following *sample loading* for 40 min (see Figure 4), the *preparative SDC run* (flow-rate = 0.2 ml/min) was continued by isocratic elution with 5% (v/v) aqueous acetic acid for 140 min, for a total run time of 180 min (2-min fractions were collected); (4) *fraction analysis* (run conditions described in Figure 4) included analysis of both the 2-min fractions (Fr. 1-90) collected during the preparative run and the components eluted isocratically off each column segment (C1-C10) following the run (as described in Figure 4). The detection wavelength was 210 nm for both analytical and preparative runs. The analytical profiles shown in panel B represent peptide components contained in various pooled fractions, together with the proportion of product, P (expressed as a percentage of total product recovered) found in each pool. Panel C shows the crossover region of product (P) and hydrophilic impurities (I) following the preparative SDC run. The peaks in the analytical profiles are all correctly proportioned.

product (C1 to C7) and 0.3% contaminated with hydrophobic impurities on column 8 (C8). Columns 9 and 10 (C9, C10), at the multi-column inlet, contained only hydrophobic impurities.

From Figure 5B, 5.4% of recovered product was found in fractions 36 and 37 contaminated with hydrophilic impurities. Figure 5C, which illustrates the separate analytical elution profiles of fractions 35 to 38, demonstrates that only one fraction (fraction 36, 0.4 ml only) comprised a major crossover fraction, where product was significantly contaminated with hydrophilic impurities. This illustrates the efficacy of the displacement process in minimizing the overlap of adjacent peptide zones under the run conditions employed.

Effect of Low Levels of Organic Modifier on Preparative SDC of LHRH (1-5). The elimination of the employment of organic modifier could have a major impact in the preparation of pharmaceutical products. The run shown in Figure 5 resulted in the elution of 88.8% of recovered product into the breakthrough fractions, illustrating that the peptide product can be successfully eluted from the column without the use of organic modifier and where the only sacrifice is an extension of run time. Further isocratic elution with 5% aqueous acetic acid would undoubtedly have removed even more of the product from the column. However, since this may lead to unfavourably long run times, the question remained whether a low level of organic modifier (methanol) added to the mobile phase (at a significantly lower concentration than that required for analytical isocratic elution) could accelerate successfully the displacement of the peptide product from the column in the SDC run, thereby maximizing yield of pure product.

Figure 6 shows the analytical profile of LHRH (1-5) resulting from a linear methanol gradient (1% methanol/min) in 0.1% aqueous TFA. The run was carried out on the multi-column set-up of Figures 3-5. Methanol is a less hydrophobic organic modifier than acetonitrile (*12*), reflected by the longer elution time (~33 min) of LHRH (1-5) in a methanol gradient (Figure 6) compared to an acetonitrile gradient (~23 min; Figures 4 and 5). The peptide is eluted at a methanol concentration of about 28% (once gradient delay time has been taken into account). The inset of Figure 6 shows the effect of different concentrations of methanol in 0.1% aqueous TFA on the isocratic elution of analytical quantities of LHRH (1-5) from the column. It is apparent that significant concentrations of methanol (≥18%) are required for isocratic elution of the peptide off the column within a reasonable time. At a concentration of 15% methanol in the mobile phase, the peptide had not been eluted from the column after 100 min in these analytical runs. The question remained whether, under high sample loads, a much lower level of methanol could be utilized to aid the sample displacement process.

In Figure 7, 100 mg of crude LHRH (1-5) has been applied to the multi-column set-up (40 min required for sample loading) and isocratic elution continued with 5% aqueous acetic acid for a further 30 min, for a run time of 70 min. Thus, at this time, the sample components have been distributed through the breakthrough fraction and column segments in a manner to that illustrated in Figure 4, i.e., the major preparative SDC separation process has now taken place. Isocratic elution was now continued with 5% methanol in 5% aqueous acetic acid for 110 min, for a total run time of 180 min (i.e., identical to that of Figure 5 for a meaningful comparison). From the results of Figure 6, this low level of methanol was deemed suitable for aiding the displacement of product from the column without simply washing product and hydrophobic impurities off the column, and thus, running the risk of contamination of purified product.

Comparing Figures 5 and 7, it is apparent that the presence of methanol has indeed moved the product further along (and aided in its displacement from)

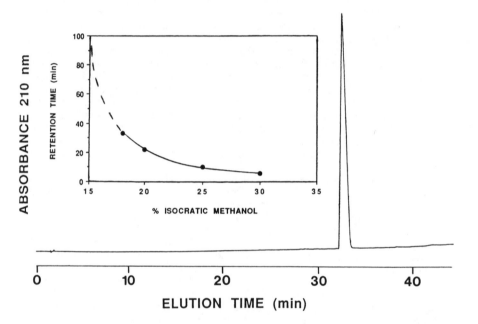

Figure 6.   Analytical RPC of LHRH (1-5). Column and HPLC instrument:
same as Figure 4.   Conditions: (1) main profile, linear AB gradient (1%
B/min) at a flow-rate of 1 ml/min, where eluent A is 0.1% aqueous TFA and
eluent B is 0.1% TFA in methanol; (2) inset profile, isocratic elution at a
flow-rate of 1 ml/min, where the eluent is 0.1% aqueous TFA containing
15%, 18%, 20%, 25% or 30% (v/v) methanol. Absorbance at 210 nm.

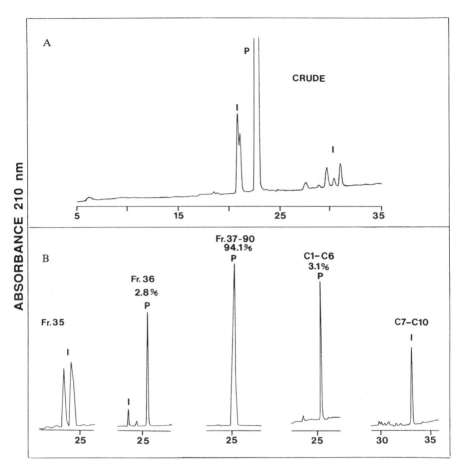

ELUTION TIME (min)

Figure 7.   Effect of low level [5% (v/v) methanol] of organic modifier as displacer on product yield following reversed-phase SDC of 100 mg of crude LHRH (1-5).   Column, HPLC instrument and sample preparation: same as Figure 4. Conditions: (1) the *analytical elution profile* (panel A) of the crude peptide (peptide product, P; impurities, I) was obtained as described in Figure 4; (2) *column equilibration*, same as Figure 4; (3) following *sample loading* for 40 min (see Figure 4), the *preparative SDC run* (flow-rate=0.2 ml/min) was carried out by isocratic elution with 5% (v/v) aqueous acetic acid for 30 min, followed by isocratic elution with 5% (v/v) methanol in 5% (v/v) aqueous acetic acid for 110 min, for a total run time of 180 min (2-min fractions were collected); (4) *fraction analysis*, same as Figure 5.   The detection wavelength was 210 nm for both analytical and preparative runs. The analytical profiles shown in panel B represent peptide components contained in various pooled fractions, together with the proportion of product, P (expressed as a percentage of total product recovered) found in each pool. The peaks in the analytical profiles are all correctly proportioned.

the column. From Figure 7B, pure LHRH (1-5) was found in fractions 37 to 90 (a volume of only 21.6 ml), representing 94.1% of recovered product, with 3.1% remaining on the column (C1-C6). From Figure 5B, pure LHRH (1-5) was found in essentially the same range of fractions (Fr. 38-90), but only represented 83.4% of recovered product, while 11.2% remained on the column (C1-C8). Overall, 96.9% of the LHRH (1-5) was eluted from the column in the presence of 5% methanol in the eluting solvent (Figure 7B), compared to 88.8% in its absence (Figure 5B).

The employment of an organic modifier in the eluting solvent was now taken a step further, both by increasing the level of methanol in the mobile phase to 10% and by its earlier introduction into the separation process.

In Figure 8, 100 mg of crude LHRH (1-5) has been applied to the multi-column set-up. As soon as sample loading was complete (after 40 min), the column was eluted isocratically with 10% methanol in 5% aqueous acetic acid, for a total run time, as before (Figures 5 and 7), of 180 min.

From Figure 8B, all of the LHRH (1-5) has been displaced from the column into the fractions, 97.3% of which was recovered pure in fractions 37-90 (a total volume of 21.6 ml). Significantly, the pure LHRH (1-5) contained in fractions 37-60 (a total volume of only 9.6 ml) represented 96% of recovered product, i.e., the best yield of purified LHRH (1-5) yet achieved, coupled with the smallest volume of pooled fractions. Since the vast majority of the product was recovered by fraction 60, the presence of 10% methanol has essentially shortened the overall run time required to complete the separation process to 120 min, in addition to accelerating the displacement of product from the column.

Comparison of Preparative SDC Approaches to Purification of LHRH (1-5). The results of Figures 3-5, 7 and 8 illustrate three variations of preparative reversed-phase SDC of peptides: (1) the standard multi-column approach, where the purified product is eluted from individual column segments following the major separation process in the absence of organic modifier (Figures 3 and 4); (2) extended isocratic elution in the absence of organic modifier, leading to displacement of all or most of the purified product from the column (Figure 5); and (3) isocratic elution in the presence of a low level of organic modifier to hasten displacement of purified product from the column, thereby decreasing overall run time (Figures 7 and 8). A summary of the yields of purified LHRH (1-5) obtained from these SDC approaches (Figures 4, 5, 7 and 8), together with details of solvent consumption, is presented in Table I.

Common advantages of all three approaches include the highly efficient utilization of stationary phase capacity, allowing the application of high sample loads (with subsequent high purified product yields) to analytical columns and employment of analytical chromatographic instrumentation. A previous report (4) highlighted this more efficient utilization of column capacity by comparing reversed-phase SDC of up to 84 mg of a five-decapeptide mixture on an analytical C_8 column (total bed length of 25 cm; 4.6 mm I.D.) with gradient elution. While 95% of total homogeneous product was obtained from the SDC approach, no more than 75% of purified product (from a much smaller sample load of 12 mg) was recovered from the gradient elution method, coupled with substantial overlap of adjacent peptide zones.

Low flow-rates characteristic of SDC ensure that solvent consumption is considerably reduced compared to traditional gradient elution methods. This is illustrated quite clearly in Table I, where 2.2 to 3.4 mg of purified LHRH (1-5) (depending on the SDC approach employed) was recovered for only a ml of total solvent consumed. Even more significant is the high level of purified product (32.8 to 81.3 mg, depending on the SDC method) obtained per ml of costly organic modifier consumed. A flow-rate considerably lower than those employed

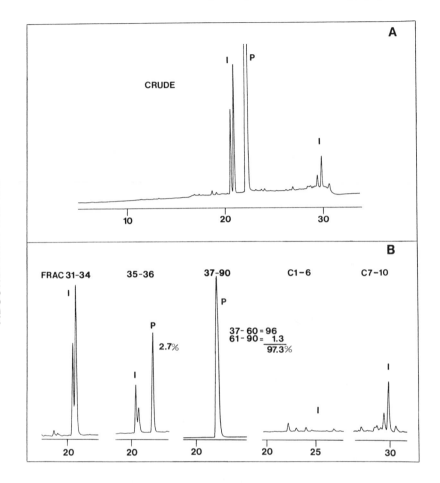

ELUTION TIME (min)

Figure 8.   Effect of low level [10% (v/v) methanol] of organic modifier as
displacer on product yield following reversed-phase SDC of 100 mg of crude
LHRH (1-5).   Column, HPLC instrument and sample preparation: same as
Figure 4.   Conditions: (1) the *analytical elution profile* (panel A) of the crude
peptide (peptide product, P; impurities, I) was obtained as described in
Figure 4; (2) *column equilibration*, same as Figure 4; (3) following *sample
loading* for 40 min (see Figure 4), the *preparative SDC run* (flow-rate = 0.2
ml/min) was continued by isocratic elution with 10% (v/v) methanol in 5%
(v/v) aqueous acetic acid for 140 min, for a total run time of 180 min (2-min
fractions were collected); (4) *fraction analysis*, same as Figure 5.   The
detection wavelength was 210 nm for both analytical and preparative runs.
The analytical profiles shown in panel B represent peptide components
contained in various pooled fractions, together with the proportion of
product, P (expressed as a percentage of total product recovered) found in
each pool.   The peaks in the analytical profiles are all correctly proportioned,
save for the pure product peak contained in fractions 37-90, which has been
scaled down 14-fold.

Table I. Comparison of Approaches to Preparative SDC of LHRH (1-5)

Preparative SDC Run of 100 mg LHRH (1-5)	Run Time (min)	Yield of Purified Product (%)	Mg/Min	Mg/Ml of Total Eluent Consumed	Mg/Ml of Total Eluent Minus Organic Modifier Consumed	Quantity of Purified LHRH (1-5) Recovered	
						Mg/Ml of Organic Modifier Consumed	Mg/Ml of Total Eluent Consumed x Run Time
Figure 4	90[a]	86.2	0.91	3.4	3.8	32.8[c]	.038
Figure 5	180	83.4[b]	0.44	2.2	2.2	-	.012
Figure 7	180	94.1[b]	0.50	2.5	2.6	81.3	.014
Figure 8	180	97.3[b]	0.51	2.6	2.8	33.0	.014

[a] Includes run time required for elution of product from individual column segments.
[b] The calculations for the SDC runs shown in Figures 5, 7 and 8 are for pure product eluted from the column during the 180 min run time.
[c] In the standard multicolumn approach, organic modifier (methanol) is only consumed for the elution of product from individual column segments.

for gradient elution also ensures high product concentration in small fraction volumes, leading to considerable savings in subsequent work-up time, e.g., for solvent removal.

Fraction analysis is also relatively straightforward, particularly for the standard multi-column approach illustrated in Figure 4, where the number of fractions to be analyzed is equal only to the number of column segments plus the breakthrough fraction. Even with the alternative approaches, where all or most pure product is completely displaced from the column, relatively few fractions need to be analyzed to locate the distribution of the desired product.

The preparative approaches described in Figures 3-5, 7 and 8 serve to illustrate the flexibility of SDC for peptide purification, where the SDC approach may be tailored to the separation goals. Thus, the standard multi-column approach (Figures 3 and 4) lends itself well to preparative purification of a crude peptide mixture for which the SDC method has not been optimized. Even under such unoptimized conditions, pure product will undoubtedly be obtained and may then be immediately employed for other applications. Any product remaining on the column segments, contaminated with impurities, is not lost and, if necessary, may be collected for a further round of SDC. The real advantages of the multi-column approach (Figure 4) are illustrated in Table 1. This approach utilized half the run time of the other methods, thus producing almost double the quantity of pure product per minute (0.91 mg/min). If one takes into consideration both run time and total eluent consumed, the multi-column approach delivers approximately three times the quantity of purified peptide (0.038 mg/ml of total eluent consumed times run time) compared to the other approaches. The runs illustrated in Figures 5, 7 and 8 represent step by step optimization of SDC for a particular peptide, where such a peptide has to be routinely purified. For example, from Figure 8, 96% of recovered pure product was obtained during the first 120 min of the run time. Thus, if the final 60 min of the run was ignored, solvent consumption (3.8 mg of product/ml of total eluent consumed; 57 mg/ml of organic modifier consumed) is reduced compared to the standard multi-column run (Figure 4; Table I). In addition, the factor linking run time and solvent consumption to amount of pure product recovered (0.032; the higher the factor, the more efficient the preparative run), is now comparable to the multi-column run. Thus, a further preparative run would only be carried out for 120 min, representing a final optimization step. In addition, the multi-segmented aspect of the column has now become secondary and, where the elution behaviour of the peptide is well characterized, SDC may be carried out on a standard analytical column. However, for non-optimized samples, the multi-column approach still remains the most successful SDC method.

It is quite apparent that the utility of SDC, in any of its variations, is not limited to the commonly employed aqueous TFA/acetonitrile mobile phase system. Where necessary, and the pharmaceutically important LHRH (1-5) is an excellent example of this, the mobile phase system may be adapted to the particular requirements of the peptide as well as its subsequent application. Which of the above approaches would be chosen for scale-up to higher sample loads on greater packing volumes is again dependent on the relative importance to the researcher or manufacturer of such factors as run time, product yield, solvent consumption, the use or otherwise of organic modifier and desired product output.

Conclusions

This paper has reviewed the early development and described current optimization of preparative reversed-phase sample displacement chromatography of peptides. The current study has, in particular, demonstrated the potential value of SDC approaches for preparative purification of peptides in the pharmaceutical and biotechnology industries. The simplicity of SDC, coupled with high yields of purified product, low consumption of HPLC solvents and minimization of potentially toxic reagents, points to a promising future for this technique.

Acknowledgments

This work was supported by the Medical Research Council of Canada and by equipment grants from the Alberta Heritage Foundation for Medical Research. We are grateful to Peptisyntha, Belgium, for donating the crude LHRH (1-5).

Literature Cited

1.   Mant, C. T.; Hodges, R. S. In *HPLC of Biological Macromolecules: Methods and Applications*; Gooding, K. M.; Regnier, F.E., Eds.; Chromatographic Science Series; Marcel Dekker, Inc.: New York, NY, USA, 1990, Vol. 51; pp. 301-332.
2.   Mant, C. T.; Zhou, N. E.; Hodges, R. S. In *Chromatography*, Part B; Heftmann, E., Ed.; Elsevier Science Publishers: Amsterdam, The Netherlands, 5th Ed.; 1991, Chapter 13.
3.   Burke, T. W. L.; Mant, C. T.; Hodges, R. S. *J. Liq. Chromatogr.* **1988**, *11*, pp. 1229-1247.
4.   Hodges, R. S.; Burke, T. W. L.; Mant, C. T. *J. Chromatogr.* **1988**, *444*, pp. 349-362.
5.   Hodges, R. S.; Burke, T. W. L.; Mant, C. T. In *Peptides: Chemistry and Biology - Proceedings of the 10th American Peptide Symposium*; Marshall, G. R., Ed.; Escom Science Publishers: Leiden, The Netherlands, 1988; pp. 226-228.
6.   Hodges, R. S.; Burke, T. W. L.; Mant, C. T. *J. Chromatogr.* **1991**, *548*, pp. 267-280.
7.   Mant, C. T.; Hodges, R. S. In *High-Performance Liquid Chromatography of Peptides and Proteins: Separation, Analysis and Conformation*; Mant, C. T.; Hodges, R. S., Eds.; CRC Press, Inc.: Boca Raton, FL, USA, 1991; pp. 793-807.
8.   Horváth, Cs.; Nahum, A.; Frenz, J. H. *J. Chromatogr.* **1981**, *218*, pp. 365-393.
9.   Antia, F. D.; Horváth, Cs. In *High-Performance Liquid Chromatography of Peptides and Proteins: Separation, Analysis and Conformation*, Mant, C. T.; Hodges, R. S., Eds.; CRC Press, Inc.: Boca Raton, FL, USA, 1991; pp. 809-821.
10. Cramer, S. M.; Horváth, Cs. *Prep. Chromatogr.* **1988**, *1*, pp. 29-49.
11. Hermodson, M.; Mahoney, W. C. *Meth. Enzymol.* **1983**, *91*, pp. 352-359.

RECEIVED December 16, 1992

Chapter 6

Displacement: Chromatographic Concentration Control

Jana Jacobson

BioWest Research, P.O. Box 135, Brisbane, CA 94005

High performance displacement chromatography (HPDC) is a practical, convenient means of controlling product concentration in protein purification to avoid extremes of concentration that can be deleterious to product quality. This study describes the purification of a complex mixture of closely related peptides by reversed phase chromatography in the HPDC mode. A major goal of the study was to compare the performance of displacement chromatography with that of gradient elution that has been the conventional mode of separating such mixtures. HPDC yielded a maximum product concentration *lower* than that obtained by gradient elution chromatography. The very high product concentrations that were observed in gradient elution separations can lead to aggregation and precipitation of a product. Such untoward effects, however, can be avoided by carrying out the chromatographic separation in the displacement mode. Hence, HPDC offered potential advantages in the ease of manufacturing for the high purity production of therapeutic compounds of interest to the pharmaceutical industry.

After scarcely two decades biotechnology has come of age, and established itself as an integral part, now, of the pharmaceutical and medical industries. The recombinant DNA technology that began as a powerful laboratory tool for understanding protein and cellular function has spawned a major industry for the large scale production of protein and peptide drugs for human and animal therapy. By many measures, such as sales figures, numbers of marketed and in-development products, variety of clinical indications targeted and stock market indicators, the biotechnology industry has matured and established itself as a significant commercial and technological player. This maturation has also necessitated a streamlining of manufacturing processes and development timelines to ensure profitability in widely differing markets and to thwart the growing competition for certain therapies. The growing health care cost burden and attendant price sensitivity for pharmaceuticals together with the relatively large dosage regimen of certain indications contribute an important element of cost-consciousness in the development of manufacturing processes in biotechnology. Competition for treatment of clinical indications with the greatest potential increases the need to rapidly develop manufacturing processes. Furthermore, regulatory considerations limit the

0097–6156/93/0529–0077$06.00/0

changes that can be made to a process, so the rapidly developed process should remain feasible as production demands increase with product launch and market expansion. These additional factors have pushed process design considerations in biotechnology farther and farther from the separation methods traditionally associated with biochemistry as practiced in the laboratory. Chromatography, in particular, has attained an outsized importance in purification processes in biotechnology, and the efficient implementation of chromatographic steps--in terms of low cost, short development time and robust operation--has received renewed engineering attention.

Chromatography is a workhorse of the bioprocessing industry for reasons that include the ubiquity and the familiarity of the technique in biochemistry laboratories, the high efficiencies and capacities attainable in packed columns, the wide variety of selectivities afforded by the varieties of available stationary phases, and the relatively simple scale-up of results obtained in the laboratory. The importance of chromatography to the industry underscores the need for readily-implemented, robust column operating modes, especially in light of the competitive factors described above. While relatively facile scale-up of separations is commonly cited as one of the advantages of process scale chromatography, in practice the operating modes adopted to optimize manufacturing processes can differ markedly from conventional--mainly analytical or micropreparative--approaches as practiced in the laboratory. These differences derive from the additional constraints on equipment capabilities, product specifications and control systems imposed at large scale.

This study describes the operational advantages of the displacement mode of chromatography (1) that has been advocated as a means for enhancing the capacity of chromatographic systems without sacrificing the high resolution obtained at relatively low column loadings in conventional elution separations (2-5). Displacement chromatography differs from elution chromatography mainly by the use of a solution of a compound with strong affinity for the stationary phase to "displace" the components of interest from the column. In operational simplicity, displacement chromatography resembles step elution chromatography, since only step changes in mobile phase composition are involved. Fundamentally, however, the displacement mode differs from elution chromatography in that the displacer adsorbs more strongly to the column surface than the feed components, unlike the eluent strength modifiers typical of elution modes. Furthermore, the adsorption properties and concentration of the displacer control both the rate and concentration at which the purified feed components exit the column. In both gradient and step elution modes of chromatography the product emerges from the column in a relatively narrow band, or peak, whose volume and concentration are a complex function of flow rate, column efficiency, eluent strength, gradient steepness, component adsorption properties and feed loading. The concentration profile of the product exiting the column in displacement chromatography differs markedly from those obtained in elution purification. The product in displacement chromatography leaves the column in a rectangular band at a concentration fixed by the displacer concentration and the relative adsorption properties of the displacer and product (1). In this study the important advantages of these features of displacement chromatography compared to the elution modes are discussed.

Experimental

Chromatography Equipment. The HPLC instrument was assembled from two Waters Model 6000A pumps (Millipore, Milford, MA USA), a Waters 720 system controller, a Rheodyne Model 7010 injector (Berkeley, CA USA) equipped with 20 μl, 2 ml and 5 ml loops and a Perkin-Elmer Model LC-75 spectrophotometer detector(Norwalk, CT USA). Fractions were collected with a Model LC-100 Haake Buchler fraction collector (Fort Lee, NJ USA). The same HPLC instrument was used for the analysis of the fractions except that injections were performed by a Waters Model 710B WISP auto-injector. The column effluent was monitored at 280 nm, and integration of chromatograms performed by the Nelson Analytical (Cupertino, CA USA) data collection system.

Materials. Synthetic luteinizing hormone releasing hormone (LHRH) was kindly provided by Dr. Suresh Kalbag of Genentech (South San Francisco, CA USA). Trifluoroacetic acid (TFA) and isopropanol (IPA) were supplied by J.T. Baker (Phillipsburg, NJ USA). Water was deionized and purified with a Milli-Q water system (Millipore) or was obtained from Black Mountain (San Carlos, CA USA). The reversed phase chromatography experiments employed the RP1 DisKit supplied by BioWest Research (Brisbane, CA USA) consisting of a 4.6 x 250 mm 5 μm column, carrier, displacer, and regenerant.

Procedures and Methods.
   Operating Steps for HPDC. After equilibrating the column with the carrier, the sample was loaded onto the column. The dual HPLC pump was programmed to generate a step gradient from the carrier to the displacer in 0.1 min, following loading of the sample. The column effluent was monitored at 280 nm, and fractions were collected over 1 or 2 minute intervals. Following the DC run, the column was regenerated by washing to remove the displacer prior to reequilibration with the carrier.
   Operating Steps for Gradient Elution. After equilibrating the column with the mobile phase, the feed was loaded onto the column in the appropriate volume. The dual pump unit was programmed to generate a linear gradient from the start conditions to the final conditions following loading of the sample. The column effluent was monitored at 280 nm, and fractions were collected over 2 minute intervals. Following the gradient run, the column was washed with the solution containing 60% IPA and 40% water with 0.1% TFA, prior to reequilibration with the starting eluent.
   Analysis of Fractions. 40 μl aliquots of collected fractions were analyzed by reversed phase HPLC using a linear gradient from 18% IPA, 0.1%TFA to 42% IPA, 0.1% TFA in water over 15 minutes. The runs were at ambient temperature and a flow rate of 0.5 ml/min. The column effluent was monitored at 280 nm, and the data integrated by the Nelson Analytical system.

Results and Discussion

Reversed phase chromatography (RPC) has developed into the separation method of choice for a wide variety of pharmaceutical compounds, and especially for small organic molecules and peptides. RPC is still gaining in importance for purification of proteins, although in certain cases the acidic hydroorganic eluents commonly employed lead to denaturation and loss of activity of proteins. Nevertheless, for those compounds able to withstand the conditions of RPC, it is one of the highest resolution separation operations available. To further explore the advantages accruing from operating RPC separations in the displacement mode, HPDC was carried out on a crude peptide mixture arising from solid-phase synthesis of LHRH. The crude mixture was used in these experiments as obtained after cleavage from the resin and therefore contained failure sequences, partially deblocked components and other compounds commonly observed in unpurified synthetic mixtures. RPC is well-established as a tool for purification of peptides, including synthetic peptides, most commonly by gradient elution chromatography. Therefore, purification of LHRH by HPDC was compared with gradient elution purification in order to examine the advantages of the former and to establish a basis for developing systems for reversed phase separations by displacement chromatography.
   Gradient elution chromatography is the standard means of purifying synthetic peptides after cleavage from the resin, and can be expected to be an important means of purifying peptides on a large scale. The approach in preparative elution chromatography is similar to analytical HPLC of peptides, and so in principle establishing the separation conditions is relatively straightforward. This is one of the main advantages of gradient

elution HPLC over alternative means of operating chromatographic process equipment: the ubiquity and familiarity of the analytical technique fosters a high comfort level with gradient elution that is not apparent toward other modes of chromatography. Figure 1A shows an analytical chromatogram of the LHRH crude mixture. Figure 1B shows a chromatogram of gradient elution purification of 40 mg of the crude LHRH mixture. Owing to the high concentration, the peak in this chromatogram is off scale and truncated. The preparative separation shown in Figure 1B was optimized by using a longer separation time to improve resolution relative to the conditions employed for analysis of the peptides. LHRH absorbs strongly at 280 nm, and closely eluting contaminants were not resolved at this loading so fractions were collected and analyzed by reversed phase HPLC. The gradient in eluent strength tends to sharpen the peak so that the product may exit the column at a higher concentration than that at which it was loaded onto the column.

The conditions of displacement separation were established by relatively small scale experiments involving injection of sub-milligram amounts of the LHRH crude mixture and following it with the displacer solution. The retention time of the feed mixture, that exited the column as a narrow band, was noted. A lack of tailing of the feed mixture band indicated that the displacer effectively sharpened the rear of the displacement train. By noting the retention time of the band, quantitative determinations of the adsorption behavior of the displacer on the reversed phase column could be made. Figure 2A shows an example of such a scouting experiment, involving the injection of 0.1 mg of the LHRH mixture onto the column. The displacement separation of the LHRH mixture was scaled up to 4 mg and 40 mg feeds, as shown in Figs. 2B and 2C. Scale up in displacement chromatography yields a longer displacement train without resulting in an increase in product concentration in the effluent.

Figure 3 shows histograms for the 40 mg elution (Figure 1B) and displacement (Figure 2C) separations. The relatively high concentration of the eluites in the elution compared to displacement separation is noteworthy. Figure 3B shows that within the displacement train, early eluting components are well separated from the main component, LHRH. The separation between LHRH and later-eluting components is poorer, suggesting that Figure 3B may not represent a fully developed isotachic displacement train. Nevertheless, most of the fractions containing authentic LHRH are of sufficiently high purity to yield a purified LHRH with high yield.

The relatively high product concentration in the effluent obtained when gradient elution is used can also effect the degree of aggregation of the molecule. The peak obtained by gradient elution has a maximum concentration that is determined by the chromatographic system and the feed loading, so that high capacity is associated with a high concentration in the product band. Essentially the concentration at the peak maximum is uncontrolled by the operator. Furthermore, since bandspreading and the effects of overloading the column result in peak broadening, the maximum concentration of the product can be many times that of the average concentration obtained after pooling the collected fractions. At this maximum concentration the product may undergo aggregation or other concentration-related changes. Concentration-dependent aggregation may be irreversible and lead to a loss in activity of the product (6). Displacement chromatography offers additional advantages compared to gradient elution separations because the product concentration is uniform in the column effluent, and can be controlled by the operating conditions to prevent the extreme concentrations encountered in overloaded gradient elution chromatography. While the peptide mixture examined here does not manifest untoward behavior related to aggregation, such behavior has been reported for proteins in reversed phase conditions (7) and may be exacerbated under high salt concentration conditions as employed in ion exchange chromatography.

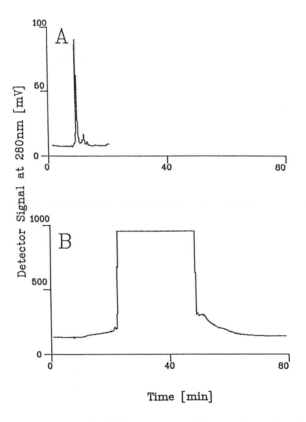

Figure 1. Gradient elution of LHRH on a reversed phase column. Gradients: (A)30-70% B in 15 minutes, (B) 0-70% B in 70 minutes where eluent A is 0.1% TFA in water and eluent B is 60% IPA and 40% eluent A. Column: BioWest RP1 4.6 x 250 mm. Operating conditions: ambient temperature, 0.5 ml/min. (A) 0.020 mg of LHRH. (B) 40 mg of LHRH.

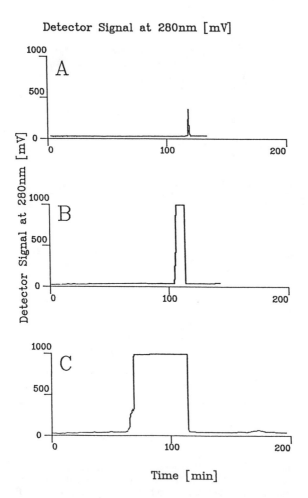

Figure 2. Displacement of LHRH on reversed phase. Displacement system: BioWest RP1 DisKit. Column: as in Figure 1. Operating conditions: ambient temperature, 0.5 ml/min. (A) 0.1 mg LHRH. (B) 4 mg LHRH. (C) 40 mg LHRH.

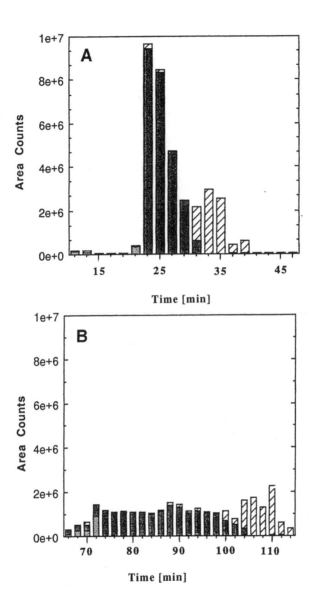

Figure 3. Histograms of elution and displacement separations. Analyses of
fractions collected during (A) elution and (B) displacement separations of 40 mg of
crude LHRH. The solid shading in the histograms represents LHRH, while the
dotted shading represents earlier eluting contaminants and the diagonal striping later
eluting contaminants in the crude mixture.

Conclusions

The displacement mode has certain inherent processing advantages compared to elution chromatography. Among these advantages are the potentially lower and controllable product concentrations in the column effluent in the displacement mode that facilitate pooling of fractions for high purity and high recovery, along with the reduced risk of product losses due to aggregation. Concentration dependent aggregation can lower the specific activity of proteins purified by gradient elution chromatography. This is because under conditions of gradient elution the peak concentration of the product can be very high and it is not easily controlled if the concentration of the target product in the feed is subject to variations. This study serves as a model, and shows that displacement chromatography may offer important benefits in protein purification, as well as peptide purification, owing to the tendency of many proteins to irreversibly aggregate under certain conditions with concomitant loss of biological activity.

The challenge of pharmaceutical use of rDNA-derived proteins is their economic recovery and purification to the specifications established by regulatory bodies. The displacement mode of chromatography offers a means of improving the operational effectiveness of high resolution separations processes and therefore can lead to important commercial and technological advantages for the biotechnology industry.

Acknowledgments

This study is based upon work supported by the award No. ISI 8960490 from the National Science Foundation. Any opinions, findings, and conclusions or recommendations expressed in this publication are those of the author and do not necessarily reflect the views of the National Science Foundation.

The author gratefully acknowledges the invaluable discussions with John Frenz and the cooperation of Hunter Tyrus Frenz in the preparation of this manuscript.

Literature Cited

1. Tiselius,A. *Ark. Kem. Mineral Geol.* **1943**, *16A*, 1.
2. Horváth, Cs. In *The Science of Chromatography*, S. Bruner, Ed., Elsevier: Amsterdam, 1985, p. 179.
3. Frenz, J.; Horváth, Cs. In *HPLC--Advances and Perspectives, Volume V*, Cs. Horváth, Ed.,; Academic Press: New York, 1988, p. 211.
4. Frenz, J.; Horváth, Cs.; *Am. Inst. Chem. Eng. J.* **1985**, *31*, 400
5. Liao, A.W.; El Rassi, Z.; Le Master, D.M.; Horváth, Cs. *Chromatographia* **1987**, *24*, 881.
6. Becker, G.W.; Bowsher, R.R.; Mackellar, W.C.; Poor, M.L.; Tackitt, P.M.; Riggin, R.M. *Biotechnol. Appl. Biochem.* **1987**, *9*, 478.
7. Mhatre, R.; Krull, I.S.; Stuting, H.H. *J. Chromatogr.* **1990**, *502*, 21-46.

RECEIVED December 15, 1992

CHROMATOGRAPHY OF GLYCOCONJUGATES

Chapter 7

Quantitative Monosaccharide Analysis of Glycoproteins

High-Performance Liquid Chromatography

R. Reid Townsend

Department of Pharmaceutical Chemistry, University of California,
San Francisco, CA 94143-0446

HPLC-based methods, which have been developed to determine
the monosaccharide composition of glycoproteins, are reviewed in
the context of approximating the exact molar ratio of individual
sugar residues to protein. Hydrolysis and methanolysis-based
methods, which have been developed for the quantitative cleavage
of all glycosidic bonds without destruction of released
monosaccharides, are discussed. Pre- and post-column
derivatization strategies for neutral and amino sugars and sialic
acids are detailed with regard to chemistry, reaction conditions,
yield and sample preparation for HPLC. The resolution and
sensitivity of the various separation and detection methods for
each class of monosaccharides (e.g. neutral, amino or anionic) are
discussed. Literature values for the monosaccharide composition
of the most commonly used model glycoprotein (bovine fetuin)
using these different methods are compared.

An accurate molar ratio of covalently-linked sugars relative to protein i)
provides the basis for further structural elucidation of glycoproteins, ii)
provides direct evidence that the polypeptide is glycosylated, iii) suggests
classes of oligosaccharide chains and iv) may serve as a measure of
production consistency for therapeutic recombinant glycoproteins. Despite
the numerous HPLC-based methods for monosaccharide analysis that have
been reported, an ideal approach has not been proven. Accurate
determination of the average number of monosaccharide residues covalently
linked to a protein generally requires quantitative, non-destructive cleavage
of glycosidic bonds, derivatization with high yields, and separation and
quantification of individual sugar residues or derivatives, all steps with
minimal losses of nanomole to sub-nanomole sample amounts. Fundamental
difficulties are due to the complex molecularity of the primary analyte--
glycoproteins. Most, if not all, glycoproteins possess multiple oligosaccharide
structures in different abundances and often with different monosaccharide
compositions. Further, the same glycoprotein (based on polypeptide

0097–6156/93/0529–0086$06.00/0

sequence) from different sources often contain different glycans (1). Finally, a single reference glycoprotein pool has not been available to investigators during the development of HPLC methods over the past 15 years. The secondary analytes, released monosaccharides, are a spectrum of chemically distinct species (Fig. 1). Some monosaccharides are closely related. For example, mannose (Man) and galactose (Gal) are 2- and 4-epimers of glucose (Glc), respectively, while others are very distinct such as Glc, a 6 carbon aldose, and N-acetylneuraminic acid (Neu5Ac), a 9 carbon carboxylated keto-sugar. The sialic acids (Fig 2), with their diversity and labile chemical groups, both expand and complicate the repertoire of methods that are needed for complete monosaccharide analysis of glycoproteins.

Overall, monosaccharide analysis of glycoproteins is a three-step process: i) monosaccharide release, ii) derivatization (when required) and iii) chromatography with quantitative detection. First, glycosidic bonds both between their monosaccharide units and between sugars and aglycons (asparagine, serine or threonine residues), must be broken without significant degradation of the component sugars. If derivatization is utilized, it must be performed with appropriate internal and external standards for each analysis. Unlike gas chromatography in which sugars are made volatile by derivatization, sugars are modified for LC both to increase their hydrophobicity [for reversed-phase-HPLC (RP-HPLC)] and to increase the sensitivity of detection. The diversity, among neutral, amino and anionic classes of monosaccharides (Fig. 1), and the isomerity within a single class has necessitated multiple derivatization chemistries and different chromatographic separations in order to analyze all sugars from glycoproteins.

In this chapter, I have detailed methods which have been developed for the HPLC-based monosaccharide analysis of glycoprotein. My intent has been to give the reader both global coverage of this area and to provide, in a single document, sufficient detail about each method so that its suitability for intended analyses can be initially assessed. The broader topic of HPLC analysis of monosaccharides from all sources is part of recent, comprehensive reviews (2-4).

Release of monosaccharides from glycoproteins

Either methanolysis, acid or enzymatic hydrolysis has been used to release intact monosaccharides by cleavage of glycosidic bonds.

Methanolysis. In the presence of excess alcohol and with HCl as a catalyst glycosidic bonds are cleaved and glycosides are formed. Fig 3 shows the four possible glycosides that are formed from the methanolysis of a β-D-glucopyranoside residue (5). The proportion of each form for a specific sugar is dependent on steric and ionic effects. Thus, during methanolysis of the oligosaccharide chains of glycoproteins, each different monosaccharide can potentially yield at least four products. For GlcNAc and GalNAc some de-N-acetylation is to be expected (6). For some sialic acids (Fig 2), de-O-acetylation is likely to occur (5).

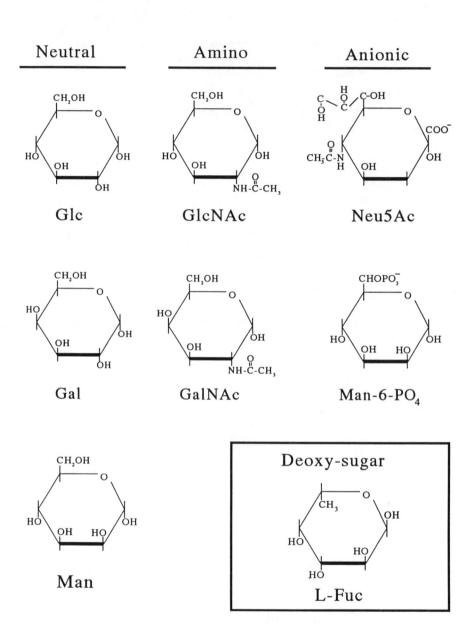

Figure 1. Monosaccharides commonly found attached to mammalian glycoproteins

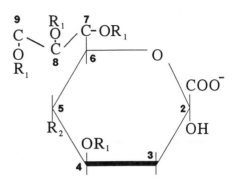

R$_1$ = H, acetyl, lactyl, methyl

R$_2$ = N-acetyl, N-glycolyl, amino, hydroxyl

Figure 2. The sialic acids

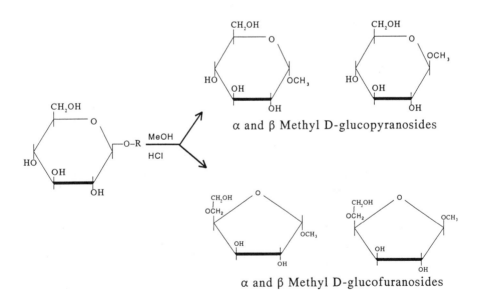

α and β Methyl D-glucopyranosides

α and β Methyl D-glucofuranosides

Figure 3. Methanolysis products of β-D-glucose

There are reports which describe HPLC methods for monosaccharide analysis of glycoproteins after methanolysis to release component sugars (6-9). Detailed methanolysis conditions [based on methods using GC (10)] have been tailored for the RP-HPLC separation and quantification of per-benzoylated methyl glycosides (6). For example, dried glycoprotein or oligosaccharide samples were treated with 1 $\underline{M}$ methanolic HCl (0.2 mL) which contained 1 $\underline{M}$ methyl acetate (water scavenger) and then heated to either 65˚C (16h) or 80˚C (4h). Prior to drying 100 μL of 80% v/v t-butanol in methanol was added to azeotropically remove water and protect against glycoside hydrolysis. Re-$\underline{N}$-acetylation of the glycosides in methanol (100 μL) was performed using pyridine (40 μL) and acetic anhydride (40 μL). $\underline{O}$-acetyl groups, which may have formed during re-$\underline{N}$-acetylation, were removed by a mild methanolysis step and the azeotropic drying procedure was repeated.

Modifications of this method (6) was recently reported (7). The re-$\underline{N}$-acetylation step was omitted and triethylamine in the methanol:t-butanol (1:4:1) was included during the drying step after methanolysis to trap HCl and minimize hydrolysis of the methyl glycosides. Analyses using narrow bore (2.1 mm) RP-HPLC enables analysis of sub-nanomole quantities of monosaccharide (8). Only methyl-1-glycosides of neutral sugars can be prepared by removing hexosamines after methanolysis (9). (In this study, the starting material were the oligosaccharides which had been released from squid cranial cartilage by β-elimination and purified from protein and peptide material using ion-exchange chromatography and ethanol precipitation.) Methanolysis was performed in sealed Pyrex glass tubes after addition of methanolic HCl (1$\underline{M}$ HCl in dry methanol) at 100°C for 30h. (Amount was not specified). After methanolysis, the samples were passed over a small (15 x 3 mm, internal diameter) Dowex 50-X8 column and washed with methanol (1 mL). The dried methanolysates were hydrolyzed in a 1% ammonia solution for 3h at 100°C. The dried sample was passed over a 3 mm I.D. ion-exchange column of 7 mm of AG 1-X8 which had been layered over 7 mm of Dowex 50-X8. The neutral methyl 1-glycosides were recovered by elution of the mixed bed ion exchange resin with 1 mL of methanol.

Acid hydrolysis. Unlike methanolysis in which one condition will release all sugars from a glycoprotein, conditions for acid-hydrolysis must be tailored for each class of monosaccharide. For example, quantitative release of amino sugars requires 4-6 N HCl at 100°C for 3-6 h for most glycoproteins, presumably for quantitative cleavage of the Asn-GlcNAc bonds. Further, amino sugars are stable under these conditions for at least 12 hours (11,12). Neutral sugars such as Man and Gal are significantly degraded under those conditions, but can be released largely intact using either 2 $\underline{M}$ HCl or 2 $\underline{M}$ TFA (12). However, the above results conflict with another report in which both Man and Gal were found to be decreased by approximately 50% using HCl (2M) for hydrolysis (13). Such differences may be related to the acid evaporation step since inclusion of Dowex 1X8 (HCO$_3^-$) reduced the losses by 20 to 30%. Most studies used 2 $\underline{M}$ TFA and 4-6 N HCl for the release of neutral and amino sugars, respectively.

Acid hydrolysis with 2 M TFA for greater than 3 hours resulted in complete de-N-acetylation of GlcNAc and GalNAc (14). If subsequent chromatography requires the N-acetylated forms for separation purposes, a method for quantitative re-N-acetylation of glucosamine and galactosamine after acid hydrolysis has been reported (15). The dried HCl hydrolysates were dissolved in a saturated solution of sodium bicarbonate (500 uL) to which 100 uL of acetic anhydride were added. The dried sample was then dissolved in "a small amount of water" and passed over a mixed ion-exchange resin consisting of Amberlite CG-120 (H$^+$ form) and Amberlite CG-400 (acetate form). The eluates (20 mL) were dried. Re-N-acetylation does enable the sample clean-up steps on cation exchange resins without losses which would occur with free amino sugars.

Sialic acids are released under much milder conditions. However, commonly used conditions (0.1 N HCl at 80°C for 30 to 90 min) result in the release of O-acetyl groups (15). Release of sialic acids with 2 M acetic acid for 1 h at 80°C was found to minimize such loss. Further these latter conditions did not result in artificial O-acetylation or destruction of sialic acids such as Neu5Ac.

Enzymatic hydrolysis. With the exception of neuraminidases with broad specificity (discussed in Ref. 15), exoglycosidases which quantitatively release monosaccharides for a set of representative glycoproteins has not been demonstrated. However, the recently described exoglycosidases from clam liver shows promise in this regard (16). A mixture of exoglycosidases which could remove all sugars from glycoproteins would eliminate many of the drawbacks discussed above for acid hydrolysis and methanolysis.

Monosaccharide Derivatives

Pre-column derivatization. Since the products from methanolysis are methyl glycosides only hydroxyl group derivatization can been used.   A detailed method for preparation of perbenzoylated methyl glycosides have been described (6). Methanolysis samples were dried overnight at room temperature followed by addition of 100 μL of reagent (10% benzoic anhydride and 5% dimethylaminopyridine in pyridine). The excess benzoic anhydride was then hydrolyzed with water. The reacted sugars were retained on a reversed-phase extraction column equilibrated in water. After rinsing (2 ml of water), the perbenzoylated monosaccharides were eluted with 2 mL of CH$_3$CN. The samples were then dried and dissolved in 20 μL of CH$_3$CN for chromatography. Recovery was found to be 80-85% using [^{14}C] glucose. Instead of RP-removal of excess hydrolyze perbenzoylation an extraction protocol has been reported (7).

Daniel reported an adaptation of a method developed for the analysis of monosaccharide alditols (17). The monosaccharides Man and Fuc were reduced with NaBH$_4$ and then perbenzoylated at 37°C for 2 h in pyridine containing 10% benzoic anhydride and 5% 4-dimethylaminopyridine. The sample was prepared for HPLC by either extraction or a by Sep-pak (Waters Assoc. Milford, MA) method. After evaporating the pyridine under nitrogen, chloroform was added and the sample was extracted sequentially with 5% sodium carbonate and then with 0.05 $\underline{M}$ HCl containing 5% NaCl. Alternatively, sample preparation using a Sep-pak was performed by first dilution of the sample with 4.5 mL of water, application to the cartridge, elution with 15 mL of water and recovery of the benzoylated alditols with 5 mL of methanol.

Acid hydrolysis of glycoproteins yields reducing sugars in which the reducing terminal can be reacted with either a fluorophore or chromophore, which serves to increase both sensitivity and hydrophobicity of monosaccharides for RP-HPLC separations. However, the simplest modification of reducing sugars is conversion to sugar alcohols (alditols). This reaction is the most straightforward, quantitative reactions for reducing sugars and has been developed into an HPLC method for quantifying the resulting monosaccharide as their alcohols (12). Since the step following acid hydrolysis is reduction with NaB[^{3}H]$_4$, the acid should be carefully removed, for example, by evaporation three times with water followed by desiccation overnight over NaOH (12). Quantitative reduction was accomplished using a 5-molar excess of NaB[^{3}H]$_4$ over 4h at 30°C in 100 μL of 50 m$\underline{M}$ NaOH. The residual reducing agent was converted to sodium borate using 200 μL of 1 $\underline{M}$ acetic acid followed by re-$\underline{N}$-acetylation of the monosaccharides by multiple additions of 10 μL of acetic anhydride. The solution was de-cationized with a 5 mL mixed-bed ion exchange resin [upper half, 2.5 mL of AG 50W-X12 (H+, 100 -200 mesh) and AG 3-X4A (OH⁻, 100-200 mesh)]. The resulting boric acid was evaporated after the addition of methanol.

Dansyl hydrazones of derivatives have been used HPLC of reducing monosaccharides (18) using a modification of a described method (19). Monosaccharides ($\sim$100 nmoles) in 16 μL of water were derivatized at 80°C for 10 min with 20 μL of an ethanolic solution of dansyl hydrazine (1% w/v) and 4 μL (10% (w/v) of trichloroacetic acid. After dilution with water (2 mL), the samples were passed over a CH$_3$CN-activated Sep-Pak in water to remove dansyl sulfonic acid and excess dansyl hydrazine. The cartridges were rinsed with 2 mL of a 10% aqueous CH$_3$CN solution. The sugar hydrazones were eluted with 40% aqueous CH$_3$CN and dried. The derivatives were reported to be stable for 2 days at 4°C. Poor derivatization was found using phenylhydrazine (13). Comparable yields were obtained using p-nitrophenylhydrazine, but the dansyl hydrazones of Gal and Glc were not separated (13). A major deficiency of this method is that 2-dexoy-2-amino sugars (e.g. glucosamine or galactosamine) do not react appreciably with dansyl hydrazine (18,19).

Monosaccharide derivatives of the azo dye, 4'-N,N-dimethylamino-4-amino-azobenzene (DAAB), have been prepared using reductive amination (20). Monosaccharides were derivatized for 10 min at 80°C with an 8-fold molar excess of DAAB, a 75-fold excess of NaCNBH$_3$ and an 8-fold excess of acetic acid. All reagents were dissolved in methanol. A 75 molar excess of pentaerythritol was present during the derivatization to scavenge borate, which is the presumed reason for low yields of DAAB-GlcNAc. The products were bound to a Sep-Pak cartridge in water and the excess DAAB was eluted with 5 mL of chloroform:hexane (1:3, v/v). The DAAB-monosaccharides were eluted with methanol (2-5 mL). Two nmoles (10 $\mu\underline{M}$)

of monosaccharide was the smallest amount reported in the derivatization. The yields, as determined by using [^{14}C]-labeled sugars, was 81-96% for neutral sugars and Fuc. However, the yields of DAAB-GlcNAc and DAAB-GalNAc were 56 and 42%, respectively. The inclusion of pentaerythritol during reductive amination caused an increase to 71% and 84%. However, the free amine counterparts, GlcN and GalN could be produced with only 10% yield and thus, a re-N-acetylation step must be included after the acid hydrolysis of glycoproteins. The DAAB-glycamines were reported to be stable in methanol and in the absence of light for greater than one year.

Pyridyl amino derivatives have been prepared from acid hydrolysates of glycoproteins (21). After acid hydrolysis, amino sugars were re-N-acetylated with 2 μL of acetic anhydride, which was added after the hydrolysates were re-dissolved in sodium bicarbonate solution. After 30 min at room temperature, 200 μL of Dowex 50W-X2 (H$^+$ form) was added, effectively to drop the pH to 3.0. The resin was then poured into a small column and washed with 5 bed volumes of water. After drying, the sample was transferred to a tapered glass tube with water and dried. Derivatization was accomplished at 100°C for 13-15 min after the addition of 5 μL of 2-aminopyridine reagent (prepared by mixing 0.5 g of 2-aminopyridine, 0.4 mL of concentrated HCl, and 11 mL of water) followed by addition of 2 μL of freshly-prepared reducing reagent (prepared by mixing 10 mg of sodium cyanoborohydride to 0.5 mL of water) and heating for 8h at 90°C. The samples were then diluted with 20 μL of water and injected onto a TSK G2000PW column (ToSoh Co., Tokyo, Japan) which was equilibrated in 20 mM ammonium acetate. The derivatized monosaccharides, found in three peaks, were pooled, dried, re-dissolved and analyzed by RP-HPLC. The yields were reported to be 70% based on an internal standard of rhamnose added after acid hydrolysis and the range of starting sugar was 10 pmol to 10 nmol. The yield of N-acetylneuraminic derivatives was only 10% and values for this sugar were not quantified.

An extraction method for pyridylamination has recently been reported which uses a pyridine-borane complex as the reducing agent (22). Saturated sodium bicarbonate and a 5% solution of acetic anhydride were added after the coupling reaction and then extracted seven time with 2 mL of benzene. This approach was performed on mucus glycoproteins (20 μg) after 4 M TFA hydrolysis and comparable ratios of sugars were found when compared to GC analysis.

Pre-column derivatization of reducing sugars with 1-phenyl-3-methyl-5-pyrazolone (PMP) has recently been described (23). The Glc product was found to contain two moles of PMP per mole of monosaccharide. Derivatives were prepared from acid hydrolysates by adding 50 μL of the PMP reagent (0.5 M in methanol) to 10 -- 500 pmol of each monosaccharide followed by 50 μL of 0.3 M sodium hydroxide. The reaction was placed at 70°C for 30 min, cooled to room temperature, neutralized with an equivalent amount of 0.1 M HCl and then dried. The samples were extracted once with equal volumes of chloroform and water and the aqueous layer was evaporated before re-dissolving and injecting into the chromatograph. Recoveries for Glc, Xyl, GlcNAc and Fuc were reported to be approximately 90%.

The primary amine of 2-deoxy-2-amino sugars can be reacted with compounds used in protein and peptide analyses. Phenylisothiocyanate (PITC) derivatives of glucosamine (GlcN), galactosamine (GalN), glucosaminitol (GlcNH$_2$) and galactosaminitol (GalNH$_2$) have been prepared for HPLC (24). Significant improvement in this approach was recently reported (11). Samples (in 20 μL) were treated with newly-prepared PITC reagent (100 μL of methanol:triethylamine:PITC, 70:10:5). Two chloroform extractions (0.8 mL each) were performed after reaction quenching using 1.5% acetic acid. Sample preparation required only 10 min. The reducing sugars were found to not form a stable product, but reacted further to form a 4-imidazoline-2-thione derivative. However, the intermediate (PTC-GlcN) could be stabilized, at least for 5h, with 1.5% acetic acid or by acid-catalyzed conversion to a stable thione derivative.

Sialic acids react with 1,2-di-amino-4,5-methylenedioxybenzene (DMB) to form fluorescent products. A sensitive HPLC-based method has been reported for the estimation of sialic acids using this reagent (25). Serum (5 μL), after mild acid hydrolysis in acetic acid and drying, was treated with 200 μL of a 7 m$\underline{M}$ DMB solution, which was prepared by dissolving DMB in 1.4 $\underline{M}$ acetic acid containing 0.75 $\underline{M}$ β-mercaptoethanol and 18 m$\underline{M}$ sodium hydrosulfite. These last two reagents were to stabilize the DMB and the sialic products, respectively. The reaction was facilitated by heating to 50°C for 3h in the dark followed by injection directly onto the chromatograph. Using radio-labeled sialic acids, a subsequent study demonstrated that the DMB reaction was not quantitative (26). This disadvantage can be overcome if appropriate standards are processed simultaneously and if the derivatization of standards is performed in a similar sample matrix (26).

Post-column derivatization. A single post-column method for the analysis of neutral and amino-containing monosaccharides in glycoproteins has been described (27). All monosaccharides or their re-$\underline{N}$-acetylated derivatives were separated isocratically on a cation exchange resin (Shodex DC-613, H$^+$ form) equilibrated with 92% aqueous CH$_3$CN at 30°C with a flow rate of 0.6 mL/min. Post-column detection was achieved by heating the sugars with 2-cyanoacetamide and monitoring the heterocyclic condensation products at 280 nm. The lower limit of detection (signal to noise = 2) for the neutral sugars was 50 -- 100 pmol while approximately 400 pmols was required for quantification of the hexosamines.

Since α-keto sugars such as Neu5Ac and Neu5Gc do not form significant uv-absorbing products with 2-cyanoacetamide a different post-column method was devised using malonitrile (28). When chromatography was performed with basic eluents, such as sodium borate buffer, pH = 9.5, then only malononitrile (0.06%, w/v) was added post-column and heated in a mixing coil (100°C). When the eluents were acidic the reagent and alkaline buffer were added separately and mixed in-line prior to the mixing coil. The fluorescent products were monitored at 430 nm with excitation at 360 nm. Detection was found to be specific for sialic acids, amino and some deoxy-sugars. The sensitivity limits were reported to be 60 pmol.

Separation, Detection and Quantification

Neutral and amino sugars. Perbenzoylated methyl glycosides (α,β-pyranosides and furanosides) were separated on a 4.6 x 150 mm RP column (3 μM Sperisorb ODS-II C$_{18}$ or 5 μM Zorbax C$_{18}$) at 1.4 mL/min (6). The separation was isocratic, 50% CH$_3$CN in water for 20 min followed by a single step to 60% CH$_3$CN. The perbenzoylated derivatives were detected using absorbance at 254 nm. Although as little as 50 pmol of sugar could be quantified in the HPLC step, approximately 1 nmole of a given sugar in the starting material was required due to small contaminating peaks introduced during sample preparation. The method was found not to be suitable for alditols of 5 carbons or greater since large amounts (~ 30%) were destroyed during methanolysis. Perbenzoylated glycosides were also separated on a Vydac 5 μm C$_{18}$ column (The Separation Group, Hesperia, CA) at 40°C which was equilibrated in 50% CH$_3$CN in water for 10 min followed by a linear gradient to 70% CH$_3$CN by 30 min (7). The flow rate was 1.5 mL/min. These workers omitted the re-$\underline{\text{N}}$-acetylation step which resulted in additional peaks for GlcNAc, GalNAc and Neu5Ac. Sensitivities in the hundreds of pmol range were reported. Reversed-phase separation of perbenzoylated monosaccharides has been performed using narrow-bore (2.1 mm) HPLC (8). The "on-column" sensitivity was reported to be in the low pmol range, although the amount of starting sample was similar to that reported using 4.6 mm columns (0.2 to 1 mg). Whether the expected enhanced sensitivity from narrow-bore separation of perbenzoylated sugars will translate into routine analysis of sub-nanomole amounts of glycoprotein was not addressed in this report. The molar ratios of the perbenzoylated methanolysis products of each sugars were similar to those previously reported (6). Methyl glycosides of Gal, Glc, Man, Xyl and Fuc (α and β forms) were separated on a CP^{TM}-MicroSpher C$_{18}$ column (Chrompack, Middelburg, The Netherlands) with elution in water. Detection was either by refractive index or post-column anthrone reaction. Two peaks were obtained for each sugar. The sensitivity of RI detection was approximately 2.5 nmoles. Other C$_{18}$ columns gave similar results if the number of theoretical plates exceeded 7000. Analysis of lactose gave the expected 1:1 ratio of Gal and Glc. Fructose methyl glycoside was hidden by the strong negative peak due to methanol. The method has only been applied to the neutral sugar analysis of a non-sulfated polysaccharide from squid cranial cartilage which was found to contain Gal, Man and Glc.

Perbenzoylated alditols were separated on a Zipax column (2.1 mm x 50 cm)(DuPont Instruments, Wilmington DL) using a gradient of dioxane (0.4 to 3% in 5 min) in hexane (17). Derivatization yields, based on recoveries of radioactive sugars, were reported to be 90 to 95%. The author noted that this method must be used in conjunction with borate paper electrophoresis since galactitol and glucitol could not be separated from mannitol using RP-HPLC.

A Pb$^+$-loaded Shodex SUGAR AP-1010 column was used to separate the alditols, GalNAcH$_2$, GlcNAcH$_2$, ManH$_2$, FucH$_2$, GalH$_2$, and Glc$_2$H, with isocratic elution of 20% ethanol and water (12). The amount of radio-labeled sugar was calculated on the basis of 2-deoxyribose as an internal standard which was added prior to the reduction step. The authors reported the methods applicability by monosaccharide analysis of bovine fetuin and recombinant erythropoietin.

Dansyl hydrazones were dissolved in approximately 20% aqueous CH_3CN and separated on a C_{18} μBondapak column (Waters Assoc., Milford, MA) (18). Dansyl hydrazones of Gal, Glc, Man and Fuc were resolved isocratically with 22% aqueous CH_3CN at 2 mL/min. The sensitivity of the method was reported to be 10 pmol using fluorescent detection (240 nm, excitation; 550 nm, emission). A subsequent study (13) described a modification necessary for automation of the dansyl hydrazine method. Gal, Glc, Man, Xyl, lyxose and Fuc were separated isocratically on a C_{18} column (10 μM, 4.6 x 250 mm) (Alltech Assoc., Deerfield, IL) equilibrated in 20% aqueous CH_3CN containing 0.01 $\underline{M}$ formic acid, 0.04 $\underline{M}$ acetic acid and 0.001 $\underline{M}$ triethylamine at a flow rate of 1 mL/min. Saturation of the column with dansylhydrazine was prevented by a column clean-up step between injections which consisted of 5 min elution with 20% CH_3CN in 80% methanol. Up to 14 successive analyses could be performed before column recycling. Routine analysis of 200-300 pmol of each sugar was reported.

DAAB monosaccharides were separated on a Supelcosil column (4.6 x 250 mm, 5 μm) (Supelco, Bellefonte, PA) equilibrated in chloroform:methanol: 0.1 M sodium tetraborate (65:25:4, v/v/v) which was adjusted to pH 3.5 with glacial acetic acid (20). The separation was isocratic at 0.75 mL/min for 15 min followed by a 2 min step gradient at 1.5 mL/min to chloroform:methanol: 0.1 $\underline{M}$ sodium acetate (65:25:4 v/v/v) at pH 3.5. The latter conditions were maintained for 13 min. Fuc, GalNAc, Man, Gal, Glc, and GlcNAc were separated, but there was significant tailing of the GlcNAc peak. The sensitivity of the method was reported to be 5 - 80 pmol.

Pyridylamininated (PA) monosaccharides were separated by RP-HPLC on a C_{18} column (Ultrasphere-ODS) (Beckman Instruments, Palo Alto, CA), which was equilibrated in 0.25 $\underline{M}$ sodium citrate buffer, pH 4.0 containing CH_3CN at 1.5 mL/min. PA sugars were detected by fluorescence (320 nm, excitation; 400 nm emission). The level of detection was 10 pmol. Except for contaminant peaks, all C_{18} columns tested gave similar chromatography with adequate separation of commonly found glycoprotein monosaccharides. In this study, sugar compositions of glycoproteins as determined by GC of trimethylsilyl derivatives and by HPLC of PA-sugars gave comparable results (21).

PMP-monosaccharides were separated isocratically (22% CH_3CN in 0.1 $\underline{M}$ phosphate buffer, pH 7.0). Separation of Man, rhamnose, GlcNAc, Glc, GalNAc, Gal, arabinose and Fuc was accomplished using a Capcell Pak C_{18} column (4.6 mm x 250 mm (Shiseido, Ginza, Chuo-ku, Tokyo) with detection at 245 nm (23). The level of sensitivity was approximately 100 pmol.

PITC derivatized GlcN, GalN and their reduced counterparts ($GlcNH_2$ and $GalNH_2$) were separated on a 5 μM C_{18} column with a flow rate of 1 mL/min (11). Eluent A was water with 0.2% of n-butylamine, phosphoric acid and tetrahydrofuran and eluent B was a 1:1 mixture of A and methanol. A linear gradient was developed over 30 min from 7% to 50% B. Regeneration and equilibration required 15 min.

Underivatized monosaccharides from acid hydrolysates of glycoproteins, both neutral and amino sugars, have been separated as their oxyanions on pellicular anion exchange resins with quaternary ammonium functional groups (CarboPac PA-1, 4.6 x 250 mm) (HPAEC) (14). Since GlcN and GalN were resolved from Fuc, Gal, Man and Glc with isocratic elution with 16 mM NaOH, re-N-acetylation was not required for analysis of acid hydrolysates. Pulsed amperometric detection (PAD) was used to monitor the eluent directly and was found to be sensitive for 10-20 pmol (S/N = 20). The molar responses of the neutral sugars was found to be approximately the same with amino sugars having a 20 -- 40% greater response. The electrochemical response of Glc was found to be linear from 50 pmol to 10 nmols (29). Monosaccharide analysis of different glycoproteins using high-pH anion-exchange chromatography with pulsed amperometric detection (HPAEC/PAD) and a post-column fluorometric method gave comparable results (14). The application of this method to a variety of glycoprotein samples from different laboratories has been reviewed (30).

Sialic acids. The diversity of functional groups on sialic acids have afforded a relatively large number of separation approaches. An excellent critical evaluation of the separation methods for sialic acids has been reported (26).

DMB derivatives of sialic acid were separated acids on a TSK-ODS 120T column (Tosoh Co., Tokyo, Japan) with CH_3CN:methanol:water (9:7:84, v/v/v) using fluorescent monitoring with excitation at 373 nm and emission at 448 nm (25). Separation of Neu5Ac, Neu5,7Ac$_2$, Neu5,9Ac$_2$, and Neu4,5Ac$_2$ (for structures, see Fig. 2) was reported, analyzed both from standards and serum. The investigators who developed the method reported a sensitivity of 100 -- 200 fmol at a S/N = 3; however, a subsequent study placed the practical sensitivity at 2.5 pmol (26).

Underivatized sialic acids have been chromatographed on amine-bonded columns in the presence of sufficient ionic strength to minimize ionic interactions (31). For example, Neu5Ac and Neu5Gc was separated using CH_3CN:water:sodium phosphate (64:26:10, v/v/v) as the eluent for a Micropak AX-5 column (300 x 4.6 mm, 5 μm) (Varian Instr., Sunnyvale, CA). The sensitivity was approximately 2 nmol by monitoring at 200 nm.

Ion-exchange chromatography using a quaternary ammonium column (No. 2633, 8 x 80 mm) (Hitachi, Danbury CT), which was equilibrated in 1.3 M borate buffer, pH 9.5 gave broad peaks for Neu5Ac and Neu5Gc and low sensitivity, approximately 2-4 nmoles (S/N = 2) (28). Ion-exclusion chromatography of these two sialic acids using a sulfonated stationary phase (Hitachi Gelpak C-620-10, H^+ form) which was equilibrated in 0.3% phosphoric acid resulted in sharper peaks and a sensitivity of 60 pmol (28). Neu5,9Ac was not separable from Neu5Ac by either method.

Underivatized sialic acids with different numbers of O-acetyl groups were separated using anion-exchange HPLC on an Aminex A-28 column (9 μm particle size) (BioRad, Richmond, CA) (32). Sialic acids were eluted isocratically in 0.75 mM sodium sulfate in 10 min. The eluent was monitored at either 215 nm or 195 nm. At the former wavelength the sensitivity was reported to be ~100 pmol. Because of the limited resolution of the method, this method was not recommended for complex biological mixtures of sialic acids (26).

Both Neu5Ac and Neu5Gc are easily separated and detected using
HPAEC/PAD (26). However, the base lability of the O-substituents of sialic
acids obviates chromatography of these sialic acids at high pH. Separation
of Neu5Ac, Neu5GcAc, Neu4,5Ac$_2$, Neu5,9Ac$_2$ and Neu5,7(8)9Ac$_3$ (for
structures, see Fig. 2) was reported using a pellicular anion exchange
(CarboPac PA-1, 4.6 x 250 mm)(Dionex Corp., Sunnyvale, CA) equilibrated
in 5 mM sodium acetate, run isocratic at 1 mL/min for 5 min, followed by a
linear gradient in 30 min to 50% 5 mM sodium acetate/50% 5 mM acetic
acid (26). The column was re-equilibrated by sequentially washing with 5
mM acetic acid for 10 min and then 5 mM sodium acetate. For PAD, post-
column base (300 mM NaOH) was added via a mixing tee at a flow rate of
0.3 - 0.4 mL/min. A routine sensitivity of 200 pmol was reported.

Comparison of reported monosaccharide compositions

Different sources of the same glycoprotein may have different types and
proportions of oligosaccharides structures. With this caveat in mind, I have
compared the values reported for the monosaccharide composition of bovine
fetuin using the methods discussed above (Table I). The expected value is
based on fetuin containing three N-linked triantennary oligosaccharide
chains, each with 5 GlcNAc, 3 Gal and 3 Man residues (33). In addition,
there is a small amount of biantennary structures have been found (34),
which contains one less lactosamine unit. There are three O-linked
disaccharide chains (35), each with one residue fo Gal and GalNAc. An
undetermined amount of a hexasaccharide (Neu5Ac→Gal(Neu5Ac→Gal→
GlcNAc)→GalNAc) has been found from fetuin (36). Estimating the
theoretical number of sialic acids is more difficult, since di, tri and
tetrasialylated species are present. Fifty-percent of the N-linked chains are
tri-sialylated and 25% are tetra-sialylated while the remainder are di-
sialylated (1). So, an estimate of 9 sialic acid residues from these structures
is reasonable. The three O-linked chains are principally monosialylated (35),
which gives a theoretical total of 12 residues of sialic acid/mol of protein.

The PITC (11), HPAEC/PAD (14), and ^{3}H-alditol (12) method gave
values (13.4 -- 16.4) for GlcN which were in reasonable agreement with the
expected value of 15 residues. All methods, with the exception of the PITC
method, apparently underestimated the GalNAc content by approximately
50%. Except for the PMP method (23), the values for Gal were
approximated those expected. However, it should be noted that the amount
of all sugars, determined by this method were low and therefore, may be
due to an incorrect protein determination, impurities or source variation.
The Man value was lower than the expected using all methods (6.4 -- 8.3
residues/mol of protein). Finally, the only method with sialic acid values
(27) indicates that the oligosaccharides of fetuin are only 60% sialylated,
which is not consistent with oligosaccharide analyses (1).

Table I   Monosaccharide Compositions of Bovine Fetuin

Reference	Fetuin Source[1]	Monosaccharide Composition (mol/mol)[2]				
		GlcN	GalN	Man	Gal	Neu5Ac
Theoretical		15	3.0	9	12	12
Honda and Suzuki (1984) (post-column)	Sigma, III	11.5	1.5	7.3	12.2	7.2
Egbert and Jones (1985) (DNS)	Sigma,III (Lot 91F-9535)	---	---	6.4	11.1	---
Takeuchi et al. (1987) (Alditol)	Sigma, III	16.4	3.5	8.3	12.6	---
Hardy et al. (1988) (HPAEC/PAD)	Gibco (Spiro)	15	1.9	7.7	13.0	---
Honda et al. (1990) (PMP)	Sigma	5.7	1.2	6.5	9.3	---
Anumula and Taylor (1991) (PITC)	Sigma	13.4	3.2	---	---	---

[1]Sigma (St. Louis, MO); Gibco (Grand Island, NY)

[2]Based on a molecular weight of 48,000 Daltons

Conclusions

Despite a plethora of different HPLC methods for the quantitative monosaccharide analysis of glycoproteins, it is still not possible to assess which method(s) best meets the criterion of providing accurate molar ratios of monosaccharide to protein. Although many studies used bovine fetuin to validate their methods, the potential variability in sugar content among preparations makes interpretation of inter-laboratory data difficult. The availability of a single pool of glycoprotein(s) and analysis by respective laboratories is essential to establish a standard analyte for the validation of methods for the quantitative monosaccharide analysis of glycoproteins.

List of Abbreviations

DAAB, 4'-N,N-dimethylamino-4-aminobenzene; DMB, 1,2-di-amino-4,5-methylenedioxybenzene; Fuc, fucose; $FucH_2$, fucitol;GC, gas chromatography; Gal, galactose; $GalH_2$, galactitol; GalNAc, N-acetylgalactosamine; GalN, galactosamine; $GalNH_2$, galactosaminitol; Glc, glucose; $GlcH_2$, glucitol; GlcNAc, N-acetyl-glucosamine; GlcN, glucosamine; $GlcNH_2$, glucosaminitol; HPAEC/PAD, high-pH anion-exchange chromatography with pulsed amperometric detection; Man, mannose; $ManH_2$, mannitol; $ManNH_2$, mannosaminitol; PITC, phenyl isothiocyanate; PA = pyridylaminated; RP-HPLC, reversed-phase high-performance liquid chromatography; S/N, signal to noise ratio; TFA, trifluoroacetic acid;

Literature Cited

1. Townsend, R. R., Hardy, M. R., Cumming, D. A., Carver, J. P. and Bendiak, B., *Anal. Biochem.*, 1989, 182, 1.
2. Honda, S. *Anal. Biochem.*, 1984, 140, 1.
3. Hicks, K. Adv. in *Adv. Carbohydr. Chem. and Biochem.*; Tipson, R. and Horton, D. Eds. Academic Press, New York, NY, 1988; Vol. 46; pp 17-77.
4. Churms, S. C. *J. Chromatogr.*, 1990, 500, 555.
5. Overend, W. G. In *The Carbohdyrates: Chemistry and Biochemistry*; Pigman, W. and Horton, D., Eds. Academic Press, New York, NY, 1972; Vol 1A; pp. 281-287.
6. Jentoft, N. *Anal. Biochem.* 1985, 148, 424.
7. Kang, E. Y. J., Coleman, R. D., Pownall, H. J., Gotto., and Yan, C.-Y. *J. Protein Chemistry*, 1990, 9, 31.
8. Gisch, D. J. and Pearson, J. D. *J. Chromatogr.*, 1988, 443, 299.
9. Hjerpe, A., Engfeldt, B., Tsegenidis, T., and Antonopoulos, C. A. *J. Chromatogr.* 1983, 259, 334.
10. Clamp, J. R. Bhatti, T. and Chambers, R. E. in *Methods of Biochemical Analysis*; D. Glick Ed. Interscience, New York, NY. Vol 19, pp. 239.
11. Anumula, K. and Taylor, P. B. *Anal. Biochem.* 1991, 197, 113.
12. Takeuchi, M., Takasaki, S., Inoue, N., and Kobata, A. *J. Chromatogr.* 1987, 207.
13. Eggert, F. M. and Jones, M. *J. Chromatogr.*, 1985, 333, 123.
14. Hardy, M. R., Townsend, R. R., and Lee, Y. C. *Anal.Biochem.* 1988, 170, 54.

15. Varki, A. and Diaz, S. *Anal. Biochem.* 1984, 137, 236-247.
16. Davidson, D. J., Fraser, M. J., and Castellino, F. J. *Biochemistry*, 1990, 5584.
17. Daniel, P. F. *J. Chromatogr.*, 1979, 176, 260.
18. Alpenfels, W. R. *Anal. Biochem.* (1981) 114, 153.
19. Avigad, G. *J. Chromatogr.*, 1977, 139, 343.
20. Rosenfelder, G., Morgelin, M., Chang, J-Y., Schonenberger, C-A., Braun, D. G. and Towbin, H. *Anal. Biochem.*, 1985, 147, 156.
21. Takemoto, H., Hase, S., and Ikenaka, T., *Anal. Biochem.* 1985, 145, 245.
22. Iwase, H., Ishii-Karakasa, I., Urata, T., Saito, T. and Hotta, K. *Anal. Biochem.* 1990, 188, 200.
23. Honda, S., Akao, E., Suzuki, S., Okuda, M., Kakehi, K., and Nakamura, J. *Anal. Biochem.* 1989, 180, 351.
24. Cheng, P.-W. *Anal. Biochem.*, 1987, 167, 265.
25. Hara, S., Yamaguchi, M., Takemori, Y., Furuhata, K., Ogura, H. and Nakamura, M., *Anal. Biochem.*, 1989, 179, 162.
26. Manzi, A. E., Diza, S., and Varki A. *Anal. Biochem.* 1990, 188, 20.
27. Honda, S. and Suzuki, S. *Anal. Biochem.*, 1984, 142, 167.
28. Honda, S., Iwase, S., Suzuki, S., and Kakehi, K. *Anal. Biochem.*, 1987, 160, 455.
29. Hardy, M. R. *Methods in Enzymol.* 1989, 179, 76.
30. Townsend, R. R. and Hardy, M. R. *Glycobiology*, 1991, 1, 139.
31. Diaz, S. and Varki, A. *Anal. Biochem.*, 1985, 150, 32.
32. Shukla, A. K. and Shauer, R. *J. Chromatogr.* 1982, 244, 81.
33. Spiro, R. G. *J. Biol. Chem.* 1962, 237, 382.
34. Townsend, R. R., Hardy, M. R., Wong, T. C. and Lee, Y. C. *Biochemistry*, 1986, 25, 5725.
35. Spiro, R. G. and Bhoyroo, V. D. *J. Biol. Chem.*, 1974, 249, 5704.
36. Edge, A. S. B. and Spiro, R. G. *J. Biol. Chem.*, 1987, 262, 16135.

RECEIVED February 9, 1993

Chapter 8

Separation of Glucose Oxidase Isozymes from *Penicillium amagasakiense* by Ion-Exchange Chromatography

Henryk M. Kalisz and Rolf D. Schmid

Department of Enzyme Technology, GBF—Gesellschaft
für Biotechnologische Forschung, Mascheroder Weg 1, DW—3300
Braunschweig, Germany

Glucose oxidase from *Penicillium amagasakiense* was purified to isoelectric homogeneity by ion-exchange chromatography on a Mono Q column using a mixed pH and salt gradient. A very flat linear gradient of 5.0-5.1% B in 40 ml using 20 mM phosphate buffer (pH 8.5) as starting buffer (A) and 50 mM acetate buffer with 0.1 M NaCl (pH 3.6) as elution buffer (B) enabled the separation of the glucose oxidase isozymes. However, the optimal sample loading and recovery of the homogeneous isoform were very low. Consequently, elution conditions were modified further. A significant increase in the optimal loading capacity of glucose oxidase on the column and a higher recovery of a single isoform were achieved with 10 mM sodium acetate buffer (pH 6) as starting buffer and 20 mM sodium acetate buffer with 0.1 M NaCl (pH 4.2) as elution buffer. No significant differences were observed in the catalytic and physical properties of the single and multiple isoforms.

Glucose oxidase (GOD) (EC 1.1.3.4) has two tightly but non-covalently bound flavin adenine dinucleotide (FAD) molecules per dimer and catalyses the oxidation of glucose to D-glucono-δ-lactone (*1*). The mechanism has been studied by several groups and a complicated reaction scheme postulated (*1-6*). The enzyme is highly specific for ß-D-glucose, although sugars such as 2-deoxy-D-glucose, D-mannose, D-galactose and D-xylose are oxidised at much lower rates (*1,7*). GOD is of considerable commercial importance (*8-11*) and is produced by several filamentous fungi, with those from *Aspergillus niger* and *Penicillium* species being commercially most important.

GOD from *Penicillium amagasakiense* is a dimer of molecular mass of 150-160 kDa and contains 11-13% carbohydrate of the high mannose type (*12-16*). The

0097—6156/93/0529—0102$06.00/0

primary structure of the *A. niger* GOD, which exhibits a high degree of similarity to the *Penicillium* enzyme (*17-21*), has recently been deduced and its gene isolated and cloned (*22,23*). Structural information about the enzyme is not available due to the lack of growth of crystals suitable for x-ray analysis. The large carbohydrate moiety, accounting for 13-24% of its molecular weight (*24,25*) is a major inhibitory factor in the crystallisation process of GOD. Cleavage of the carbohydrate moiety of *A. niger* GOD enabled the growth of crystals suitable for x-ray diffraction analysis (*24*). Deglycosylation of the *P. amagasakiense* GOD did not effect a reduction in the isoelectric heterogeneity of the enzyme (Hendle, J., Kalisz, H.M., Schomburg, D., Schmid, R.D., unpublished data), which may have hindered the crystallisation process. Hence, purification to isoelectric homogeneity prior to deglycosylation was necessary (*26*).

This paper demonstrates the ability of ion-exchange chromatography to resolve *P. amagasakiense* GOD isoforms to isoelectric homogeneity. Very flat mixed pH and salt gradients were used to separate the isoforms. Purification conditions were optimised to increase sample loading and yield. The properties of the single GOD isozyme were compared with those of the heterogeneous preparation.

Experimental

Materials.   GOD from *P. amagasakiense* was purchased from Nagase Biochemicals Ltd. (Japan). All chemicals were from Merck (Darmstadt, F.R.G.). PhastGel polyacrylamide gels were obtained from Pharmacia (Uppsala, Sweden).

Enzyme assay.   GOD activity was assayed at 420 nm as described previously (*25*) using 2,2'-azino-di-[3-ethylbenzthiazoline-6-sulphonic acid] (ABTS) as the dye and 0.1 M glucose as the substrate. Assays were performed in 0.1 M acetate buffer, pH 6, at 25°C under oxygen saturation.

Protein was determined by the method of Bradford (*27*) using Coomassie Brilliant Blue G reagent (Biorad) with bovine serum albumin as the standard.

Electrophoresis.   Polyacrylamide gel electrophoresis (PAGE) was performed either in the presence (dissociating conditions) or absence (non-dissociating conditions) of sodium dodecyl sulphate (SDS) on 10-15% or 8-25% gradient gels, respectively, using the Pharmacia Phast System (Pharmacia LKB, Uppsala, Sweden) according to the manufacturer's instructions (*28*). Isoelectric focusing was performed in the pH range of 4.0-6.5 according to Olsson et al. (*29*). Electrophoretic titration was performed in the pH range of 3.0-9.0 as described by Jacobson & Skoog (*30*). Gels were silver-stained by the method of Butcher & Tomkins (*31*).

Purification.   Purification was performed with a Pharmacia FPLC unit equipped with two P-500 pumps, an LCC-500 controller and an LKB 2238 Uvicord SII ultraviolet monitor fitted with a 280 nm filter. Chromatograms were recorded with an LKB 2210 two-channel recorder. Samples were collected with a FRAC-100 fraction collector. GOD was dissolved in and dialysed against the start buffer and applied to a Mono Q HR 5/5 column preequilibrated with the appropriate buffer. All samples and buffers were filtered through 0.22 μm Millex-GV$_{13}$ (Millipore) or Cellulose Acetate (Sartorius) filters before use. GOD containing fractions were pooled, desalted and concentrated as described previously (25). Sample purity and homogeneity were assessed electrophoretically. Unless otherwise stated a flow rate of 2 ml/min was used.

Compositional Analysis.   Amino acid composition was determined with a Biotronic LC5001 amino acid analyser (Munich, FRG) following hydrolysis in 6 M HCl at 105°C for 24-72 h. Carbohydrate contents were estimated with a Carlo Erba Series GC gas chromatograph (Hofheim, FRG) according to Chaplin (32).

Results and Discussion

Glucose oxidase from *P. amagasakiense* exists in multiple isoelectric forms with isoelectric points between pH 4.3-4.5 (Hendle, J., Kalisz, H.M., Schomburg, D., Schmid, R.D., unpublished data). Ion-exchange chromatography on a Mono Q column was used to purify GOD from *P. amagasakiense* to isoelectric homogeneity (26). The influence of the steepness of the salt gradient on GOD elution at different pH values was initially tested. The resolution of the GOD isozymes was poor under all conditions tested.

The resolution was improved by using a flat mixed pH and salt gradient, with 20 mM phosphate buffer (pH 8.5) as the starting buffer (A) and 100 mM acetate buffer with 200 mM NaCl (pH 3.6) as the elution buffer (B) (Figure 1).

Further improvement in the isozyme resolution was obtained by a flattening of the linear gradient and a twofold dilution of buffer B (results not shown). Subsequent modifications of the gradient slope dramatically affected the elution profile of the enzyme. Optimal resolution was achieved with 50 mM acetate buffer with 100 mM NaCl (pH 3.6) as buffer B and a linear gradient of 5.0-5.1% B in 40 ml at a flow rate of 1 ml/min (Figure 2A). The isoelectric points of each of the 6 peaks are shown under the corresponding peak in Figure 2B. Each of the three GOD peaks eluted with the flat gradient of 5.0-5.1% B comprised a maximum of two isoforms (Figure 2B). Moreover, each of these peaks exhibited a homogeneous profile in the pH range 3-9 on an electrophoretic titration curve (Figure 3). Peaks 4-6, eluted at 10%, 30% and 100% B, were heterogeneous with respect to their isoelectric population and titration curve profiles.

Crystallisation attempts with each of the 6 GOD peaks were performed as described in (33). Crystals were obtained only with the enzyme eluted as peak 1. GOD eluted as peaks 2-6, including the relatively homogeneous peaks 2 and 3, could not be crystallised. Thus, only peak 1 was used further. However, due to the nature of the gradient both the recovery of the homogeneous isoform (10%) and the optimal loading capacity (1 mg) of GOD on the Mono Q column were low. Thus, the purification procedure had to be repeatedly re-run in order to obtain sufficient material for the crystallisation. Hence, it was time-consuming and expensive. Consequently, further modifications of the buffer composition and elution conditions were attempted.

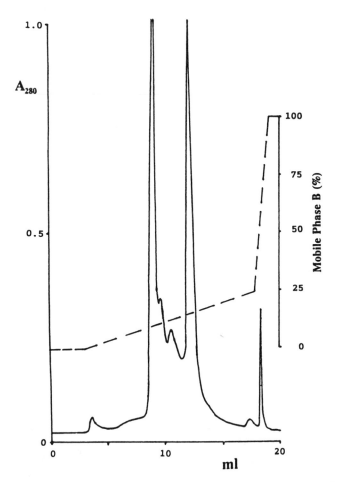

Figure 1. Influence of a mixed pH and salt gradient development time on the separation of *Penicillium amagasakiense* GOD isoforms. GOD was applied to a Mono Q column in 20 mM potassium phosphate buffer (pH 8.5) (buffer A) and eluted with 100 mM sodium acetate buffer-0.2 M NaCl (pH 3.6) (buffer B) with a gradient of 0-25% B in 15 mL. (Reproduced with permission from ref. 26. Copyright 1990 Elsevier.)

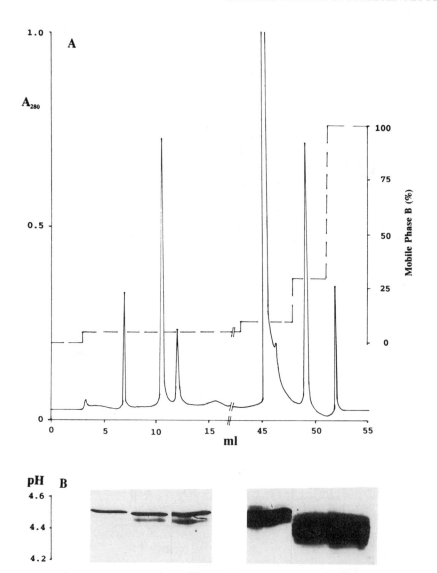

Figure 2. Separation of *Penicillium amagasakiense* GOD isoforms on a Mono Q column (A) and by isoelectric focusing (B). GOD was applied to a Mono Q column in 20 mM potassium phosphate buffer (pH 8.5) (buffer A) and eluted with 50 mM sodium acetate buffer-0.1 M NaCl (pH 3.6) (buffer B) with a linear gradient of 5.0-5.1% B in 40 mL, then in stepwise increments at 10% B, 30% B and 100% B. The isoelectric points of each of the corresponding peaks were determined by isoelectric focusing in the range of pH 4.0-6.5 as described in Experimental. (Reproduced with permission from ref. 26. Copyright 1990 Elsevier.)

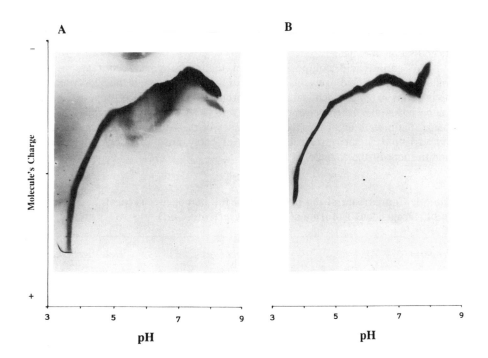

Figure 3. Electrophoretic titration curve of *Penicillium amagasakiense* GOD (A) before and (B) after purification (peak 1) under the conditions described in Figure 2. (Reproduced with permission from ref. 26. Copyright 1990 Elsevier.)

The difference in the pH of the starting and eluting buffers was assumed to be the main reason for the necessity of the flat gradient required for the resolution of the homogeneous GOD isoform. The difference in the pH of buffers A and B was gradually reduced and a number of elution conditions were tested. The narrower the pH difference of buffers A and B, the steeper was the gradient at which the homogeneous GOD isoform could be separated. Optimal resolution was achieved using 10 mM sodium acetate buffer (pH 6) as buffer A and 20 mM sodium acetate buffer with 100 mM NaCl (pH 4.2) as buffer B, and a linear gradient of 10-18% B in 16 ml (Figure 4). Under these conditions both the recovery of the homogeneous isoform (60%) and the optimal loading capacity of the Mono Q column (10 mg) were significantly increased. The peak eluting at 100% B comprised a heterogeneous isoform population and was not used for further crystallisation attempts.

The amino acid compositions and the carbohydrate contents of the homogeneous and heterogeneous isoforms of the enzyme were identical. Moreover, the isoforms exhibited identical catalytic properties. Some of the properties of the homogeneous and heterogeneous isoforms of GOD are summarised in Table I. Hence, the protein and carbohydrate moieties of the GOD isoforms are identical and variations in their isoelectric points probably arise from post-translational modifications of the enzyme.

Table I. Comparison of the properties of the homogeneous (peak 1, Figure 2) and heterogeneous isoforms of *P. amagasakiense* GOD

Property	Homogeneous	Heterogeneous
pI	4.51	4.46-4.32
M_r (kDa)	167	167
Amino acid content	1220	1222
Carbohydrate content (%)	13.1	13.2
K_m (mM)	5.2	5.2
k_{cat} (s^{-1})	1475	1447
pH optimum	4.5-5.5	4.5-5.5
pH stability	5-7	5-7
T optimum (°C)	40-50	40-50
T stability (°C)	50	50

The ability of ion-exchange chromatography to resolve *P. amagasakiense* GOD isoforms to isoelectric homogeneity with a very intricate mixed pH and salt gradient has been demonstrated. The homogeneous *P. amagasakiense* GOD isoform has been successfully crystallised and the preliminary structure solved (*33*).

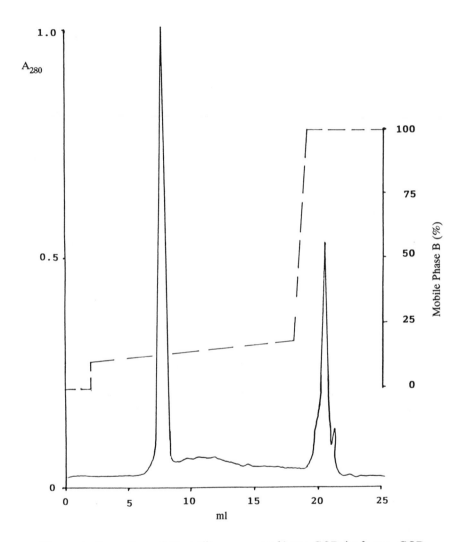

Figure 4. Separation of *Penicillium amagasakiense* GOD isoforms. GOD was applied to a Mono Q column in 10 mM sodium acetate buffer (pH 6) and eluted with 20 mM sodium acetate buffer-0.1 M NaCl (pH 4.2) (buffer B) with a linear gradient of 10-18% B in 16 mL.

Acknowledgments

We would like to thank Claudia Stein-Kemmesies for her invaluable technical assistance and Dr. Manfred Nimtz for analysis of the protein-bound carbohydrate content.

Literature Cited

1. Bright, H.J.; Porter, D.J.T. In *Enzymes*; Boyer, P.D., Ed.; 3rd ed., Academic Press, New York, 1975, Vol. 12; p. 421.
2. Nakamura, S.; Ogura, Y. *J. Biochem.* **1962,** *52,* 214.
3. Nakamura, S.; Ogura, Y. *J. Biochem.* **1968,** *63,* 308.
4. Gibson, Q.H.; Swoboda, B.E.P.; Massey, V. *J. Biol. Chem.* **1964,** *239,* 3927.
5. Bright, H.J.; Gibson, Q.H. *J. Biol. Chem.* **1967,** *242,* 994.
6. Duke, F.R.; Weibel, M.; Page, D.S.; Bulgrin, V.G.; Luthy, J. *J. Am. Chem. Soc.* **1969,** *91,* 3904.
7. Keilin, D.; Hartree, E.F. *Biochem. J.* **1952,** *50* 331.
8. Crueger, A.; Crueger, W. In *Biotechnology;* Rehm, H.-J.; Reed, G., Eds.; Verlag Chemie, Weinheim, 1984, Vol. 6a; p. 421.
9. Röhr, M.; Kubicek, C.P.; Kominek, J. In *Biotechnology;* Rehm, H.-J.; Reed, G., Eds.; Verlag Chemie, Weinheim, 1983, Vol. 3; p. 455.
10. *Biosensors. Fundamentals and Applications;* Turner, A.P.F.; Karube, I.; Wilson, G.S., Eds; Oxford University Press, Oxford, 1987.
11. Schmid, R.D.; Karube, I. In *Biotechnology;* Rehm, H.-J.; Reed, G., Eds; Verlag Chemie, Weinheim, 1988, Vol. 6b; p. 317.
12. Kusai, K.; Sekuzu, I.; Hagihara, B.; Okunuki, K.; Yamauchi, S.; Nakai, M. *Biochim. Biophys. Acta* **1960,** *40, 555.*
13. Nakamura, S.; Fujiki, S. *J. Biochem.* **1968,** *63,* 51.
14. Yoshimura, T. Isemura, T. *J. Biochem.* **1971,** *69,* 839.
15. Yoshimura, T. Isemura, T. *J. Biochem.* **1971,** *69,* 969.
16. Hayashi, S.; Nakamura, S. *Biochim. Biophys. Acta* **1976,** *438,* 37.
17. Swoboda, B.E.P.; Massey, V. *J. Biol. Chem.* **1965,** *240,* 2209.
18. Tsuge, H.; Natsuaki, O.; Ohashi, K. *J. Biochem.* **1975,** *78,* 835.
19. O'Malley, J.J.; Weaver, J.L. *Biochem.* **1972,** *11,* 3527.
20. Hayashi, S.; Nakamura, S. *Biochim. Biophys. Acta* **1981,** *657,* 40.
21. Jones, M.N.; Manley, P.; Wilkinson, A. *Biochem. J.* **1982,** *203,* 285.
22. Kriechbaum, M.; Heilmann, H.J.; Wientjes, F.J.; Hahn, M.; Jany, K.-D.; Gassen, H.G.; Sharif, F.; Alaeddinoglu, G. *FEBS Lett.* 255 (1989) 63.
23. Frederick, K.R.; Tung, J.; Emerick, R.S.; Masiarz, F.R.; Chamberlain, S.H.; Vasavada, A.; Rosenberg, S.; Chakraborty, S.; Schopfer, L.M.; Massey, V. *J. Biol. Chem.* **1990,** *265,* 3793.
24. Kalisz, H.M.; Hecht, H.J.; Schomburg, D.; Schmid, R.D. *J. Mol. Biol.* **1990,** *213,* 207.
25. Kalisz, H.M.; Hecht, H.J.; Schomburg, D.; Schmid, R.D. *Biochim. Biophys. Acta* **1991,** *1080,* 138.

26.   Kalisz, H.M.; Hendle, J.; Schmid, R.D. *J. Chromatog.* **1990,** *521,* 245.

27.   Bradford, M.M. *Anal. Biochem.* **1976,** *72,* 248.

28.   Pharmacia Phast System™ System Guide, Pharmacia Laboratory Separation Division, S-75182, Uppsala, Sweden.

29.   Olsson, I.; Axio-Fredriksson, U.; Degerman, M.; Olsson, B. *Electrophoresis,* **1988,** *9,* 16.

30.   Jacobson, G.; Skoog, B. In *Electrophoresis '86;* Dunn, M.J., Ed.; VCH Verlagsgesellschaft, Weinheim, 1986; p. 516.

31.   Butcher, L.A.; Tomkins, J.K. *Anal. Biochem.* **1985,** *148,* 384.

32.   Chaplin, M.F. *Anal. Biochem.* **1982,** *123,* 336.

33.   Hendle, J.; Hecht, H.J.; Kalisz, H.M.; Schomburg, D.; Schmid, R.D. *J. Mol. Biol.* **1992,** *223,* 1167.

RECEIVED October 14, 1992

Chapter 9

Analysis of Microsomal Cytochrome P-450 Patterns

Fast Protein Liquid Chromatography with Ion-Exchange and Immobilized Metal Affinity Stationary Phases

P. H. Roos

Institute of Physiological Chemistry I, Ruhr–University Bochum, W–4630, Bochum, Germany

Xenobiotically and endobiotically caused variations in the profile of the multiple liver microsomal cytochrome P450–isozymes result in altered metabolic activities towards drugs, carcinogens and endogenous steroids or eicosanoids. Our aim is the comprehensive analysis of these alterations in rat liver. This task, however, is difficult because of the large number of biochemically similar P450–isozymes. Therefore, our approach includes the combination of different analytical methods starting with fractionation of the microsomal P450–species by high–resolution ion–exchange and immobilized metal affinity fast protein liquid chromatography (FPLC). Here we describe optimization of both methods for P450–fractionation. Chromatographic resolution can be markedly increased by application of combined segmented/stepwise gradients for either method and is monitored by the elution profiles, SDS–PAGE patterns and immunoblots of individual fractions.

Much work has been done on the liquid chromatographic separation of microsomal cytochromes P450 on a preparative scale *(1)*. Usually, conventional low–pressure techniques are used including many consecutive chromatographic steps *(2)*. The aim of P450–purification is the enzymatic and biophysical characterization of the isolated species. Often, the price of the desired high purity of the preparation is the low yield *(e.g., 3)*. With some exceptions, only a few chromatographic principles are applied to P450–separation, i.e. primary fractionation on aminooctyl–Sepharose (a combined hydrophobic interaction / weak anion exchange method) and subsequent chromatography on ion–exchange resins and hydroxyapatite *(1)*. The application of high–resolution HPLC methods for P450–fractionation, introduced by Kotake et al. *(4)*, is limited and has been always carried out in combination with low–pressure methods *(5-7)*. The high number of necessary chromatographic steps and the resulting low yields are caused by the great number of biochemically similar P450–species present in the microsomal fraction. By recent estimates, more than 60 different P450–isozymes may be potentially present in a mammalian species *(8)*. Due to the inducibility of many isozymes *(9)*, the microsomal P450–pattern changes in response to the body's external and internal (hormonal) environment. As a consequence, the metabolic activities towards xenobiotics, i.e. drugs and environmental toxins, and endogenous compounds, such as steroids and eicosanoids, vary considerably *(10)*.

0097–6156/93/0529–0112$06.00/0

The knowledge of the individual P450-pattern allows the evaluation of xenobioti-cally caused effects, and identification of the chemical nature of the effector com-pound, and might be important in the design of therapeutic measures (11,12). Pattern analysis, however, is difficult because of the large number of P450-species which are similar in molecular weight (1), enzymatic activity (13), spectroscopic properties (14) and immunochemical reactivity (15). Therefore, recognition and quantitation of indivi-dual, closely related P450-isozymes and comprehensive pattern analysis require a net-work of different methods with chromatographic fractionation of the isozymes as the primary step.

The requirements for analytical chromatographic separations are often very diffe-rent from those for preparative methods. The main goals for analytical procedures are high resolution and reproducibility of each chromatographic step. For this purpose FPLC-methods based on ion-exchange chromatography and immobilized metal affi-nity chromatography (IMAC) were optimized. Because sample preparation and the chromatographic procedure itself might lead to partial isozyme loss or damage, analy-ses of unfractionated microsomes by enzymatic, spectroscopic and immunochemical methods are routinely applied in addition. However, results concerning this part of our work are not presented here.

Sample Preparation for Analytical Fractionation of Microsomal Cytochrome P450-Isozymes

Selection of a Suitable Detergent. Separation of membrane-bound cytochromes P450 by column chromatography requires their solubilization by detergents. Most com-monly, sodium cholate is used for this purpose. However, because cholate interferes with the intended chromatographic separation on the strong anion exchanger Mono Q (16,17), we had to select another detergent. As a criterion for suitability of a detergent the percentage of spectroscopically intact cytochrome P450 was determined after solu-bilization and removal of the unsolubilized material by ultracentrifugation (16,17). Ba-sed on these studies, we have chosen 0.8% Lubrol PX for routine solubilization with a 96% recovery efficiency of the solubilized P450. This value is similar to the recovery efficiency reported for cholate (17,18). Cationic detergents lead to irreversible damage of P450 and were excluded from further studies (16).

Besides high solubilization efficiency, the detergent used has to fulfill two further criteria: it has to be compatible with and suitable for the chromatographic procedure, and the enzymatic activity of the P450-species has to be maintained. As has been shown, Lubrol PX fulfills these criteria (17). An additional advantage of Lubrol is that it does not absorb at 280 nm, so that protein monitoring at this wavelength is possible during chromatography. The stabilization of P450 during membrane disintegration and subsequent chromatography requires the presence of glycerol (24) and detergent (25).

Usually, a combination of cholate and a non-ionic detergent, e.g., Lubrol PX, Emulgen 911 or Renex (19,20,26), is used to obtain homogeneous P450 preparations (1). However, Dutton et al. (27) have shown that this mixture could be successfully replaced by the zwitterionic detergent CHAPS. Compared to the non-ionic detergents, an advantage of CHAPS is the low inhibitory effect on P450-dependent enzymatic ac-tivities. While cholate and probably also CHAPS function as lipid substitutes (28,29), non-ionic detergents disturb the interaction of cytochromes P450 with the NADPH-cytochrome P450-reductase (28-30). After partial removal of inhibitory Lubrol by adsorption on Biobeads we are measuring P450-activities in a reductase-independent manner using cumene hydroperoxide as an artificial oxygen donor (17,31).

Undesired Effects of Sample Pretreatment. For the purpose of preparative P450-se-paration, the next steps after solubilization commonly used by others are either pre-fractionation by poly(ethylene glycol) (PEG) (19) or chromatography on aminooctyl-Sepharose (20) in order to separate the majority of cytochromes P450 from the bulk of other proteins. Neither steps are useful for analytical separations as intended here because they lead to a pronounced loss of cytochrome P450. In our studies, we routinely recover about 50 % of the applied P450 in the combined eluted fractions from the aminooctyl-Sepharose column, in accordance with results reported in the literature *(5,21-23)*. Besides cytochrome P420 (the enzymatically inactive form of cytochrome P450), the pass-through fraction contains considerable amounts of intact P450. Because of its low specific P450-content, this pass-through fraction is usually omitted for preparative purposes. Nevertheless this fraction might include P450 forms not present in the P450-enriched eluate. The same is true for PEG-fractionation leading to an initial loss of about 25 to 40% P450 *(5,6,21-23)*. Therefore, such prefractionations are not useful for analytical separations. Instead, we are using detergent- solubilized microsomes as samples for this purpose.

Analytical Ion-Exchange FPLC of Microsomal Cytochromes P450.

High-resolution chromatographic techniques, such as HPLC and FPLC, have been applied only infrequently for the separation of P450. On a preparative scale, HPLC is sometimes used as a final step for P450 purification *(3-5,21-23,32)*. As an analytical tool in P450 research, the use of HPLC is reported in only a few studies *(4,33-35)*. FPLC has been applied for this purpose only by Sakaki et al. *(36)* to show the distinctness of purified P450-isozymes and by Kastner & Schulz *(37)* to differentiate between P450-patterns in liver microsomes of marmoset monkeys treated with various inducers. Optimization of ion-exchange FPLC for the purposes *(16,17)* outlined in the Introduction is described in the following sections.

Optimization of the Chromatographic Resolution. To optimize the resolution of P450-separation on the FPLC ion-exchange resins Mono Q and Mono S the following parameters were examined: type of detergent in the eluent buffer, pH, gradient form and column load. If not otherwise stated, the following buffers have been used for chromatography: Mono Q, equilibration (Aq): 20 mM Tris-HCl; pH 7.7, 20 % glycerol, 1 mM EDTA, 1 mM DTE, 0.2 mM PMSF, 0.2 % Lubrol PX. Mono Q, elution (Bq): buffer Aq + 1 M NaCl. Mono S, equilibration (As): same composition as Aq but with 20 mM MOPS, pH 7.0 instead of Tris-HCl. Mono S, elution (Bs): buffer As + 1 M NaCl.

Experiments with sodium cholate (instead of Lubrol PX) showed that this detergent leads to reduced binding capacity of the column and irreproducible elution profiles with respect to retention times and peak heights (results not shown here). The reason is the strong binding of cholate to the anion-exchange resin. Therefore, cholate is not suitable for analytical fractionations on this type of column. These problems were circumvented by the use of Lubrol PX, resulting in highly reproducible elution profiles. Mean retention times of several runs are listed in Table I. Peak No. 4 may

Table I. Reproducibility of P450-separation by anion-exchange FPLC on Mono Q evaluated by retention times of individual peaks

Peak-No.	1	3	4	5	6	7
Mean	9.50	20.41	29.01	35.16	41.67	49.32
s.d.	1.68	0.46	0.58	0.20	0.46	0.84
n	8	6	8	6	6	5

Peak numbers correspond to those given in Figure 2. Values give the Means and standard deviations (s.d.) of the retention times (min). n: number of determinations. The retention times were taken from chromatographic runs using solubilized liver microsomes of untreated rats or rats treated with phenobarbital, ß-naphthoflavone or hexachlorobenzene.

function as an internal standard for retention times. Besides some cytochrome P450, this peak mainly contains cytochrome b_5 (*17*), a component of the microsomal monooxygenase system showing no isozymic variation. When chromatographed on Mono Q, the solubilized microsomes which were first prefractionated on aminooctyl-Sepharose, i.e., freed of cytochrome b_5, this peak was considerably reduced in size.

Effect of the Gradient Shape and Sample Load. Initially, simple linear NaCl-gradients were used for the elution of P450. The resulting low chromatographic resolution could be markedly increased, however, by the use of stepped and segemented gradients (Figures 1, 2 and (*17*)). Using the optimized gradient system, eight P450-containing fractions including the pass-through fraction were obtained with partial overlap of some peaks detected by continuous monitoring of heme proteins at 417 nm (Figure 2). By using the same column geometry and elution schedule, the resolution decreases with higher sample load (*17*) (Figure 3). This results in a shift of the apparent retention times, especially for peaks 4 and 5 (Table II).

Table II. Influence of sample load on the retention time in ion exchange FPLC on Mono Q

Microsomal sample		Retention times (min) of peak		
Inducer	nmol P450	4	5a	5b
PB	0.92	28.00	s.s.	33.75
PB	6.55	30.85	32.65	34.35
ßNF	0.92	28.72	31.60	35.17
ßNF	5.77	30.59	34.50	35.36
HCB	0.92	28.93	s.s.	35.02
HCB	4.82	30.00	32.43	34.93

Peaks 4, 5a and 5b correspond to the chromatograms shown in Figure 3. s.s. = slight shoulder.

For analytical fractionations on Mono Q HR5/5 with a bed volume of 1 ml, the recommended sample input is 1 to 2 nmol cytochrome P450 with solubilized microsomes as the starting material. For the fractionation of microsomal P450s from marmoset monkeys, Kastner & Schulz (*37*) used sample sizes in the same range. Typical elution profiles obtained with solubilized liver microsomes of rats treated with various inducers are shown in Figures 2 and 3. SDS-PAGE patterns of individual fractions are given in Figure 4 . Further analyses of the fractions include quantitative spectroscopic determination of P450 (Table III) and detection of enzymatic activities (*17*).

Identification of isozymes and conclusions. The following conclusions can be drawn from these analyses: individual peak fractions may contain more than one P450-species requiring further resolution and several fractions exhibit similar enzymatic activi ties (*17*) and SDS-PAGE-patterns (Figure 4). This may be explained by the presence of closely related isozyme forms in the separated fractions. We want to discuss this point in more detail. In accord with our findings (*17*), Funae & Imaoka (*21*) detected testosterone 6ß-hydroxylase activity in 5 out of 6 fractions obtained by anion-exchange HPLC. This activity is expressed by P450 3A-isozymes, four forms of which could be chromatographically resolved by Nagata et al. (*38*).

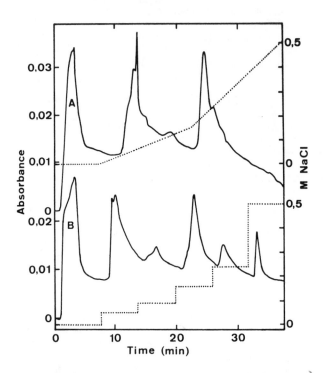

Figure 1.   Effect of the gradient form on the resolution of P450–species by anion–exchange FPLC on Mono Q HR5/5. Samples: Lubrol–solubilized liver microsomes of male rats concomitantly treated with phenobarbital and ß–naph-thoflavone with a P450–content of 1.25 nmol. Flow rate: 1 ml/min. Solid line: absorbance at 417 nm; broken line: NaCl–gradient.

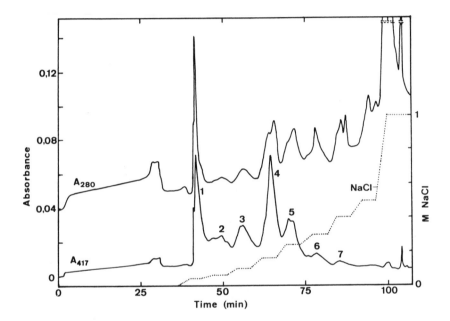

Figure 2. Fractionation of solubilized liver microsomes by optimized anion-exchange FPLC. Column: Mono Q HR5/5. Sample: Lubrol-solubilized liver microsomes of rats treated with dexamethasone, P450-content: 4.7 nmol. Flow rate: 1 ml/min. Absorbance at 417 nm and 280 nm is shown as indicated. Dotted line: NaCl-gradient.

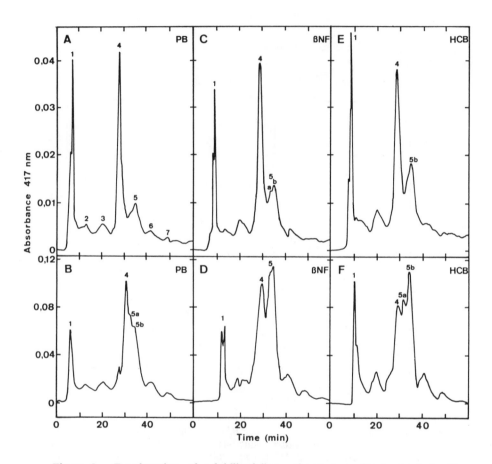

Figure 3. Fractionation of solubilized liver microsomes of rats treated with different inducers: Effect of sample load on P450-resolution. Fractionation was done by the optimized method with the NaCl-gradient shown in Figure 2. Column: Mono Q HR5/5. Samples: Lubrol-solubilized liver microsomes of rats treated with phenobarbital (A,B), ß-naphthoflavone (C,D) and hexachlorobenzene (E,F). P450-content of the samples: A, 0.93 nmol; B, 6.55 nmol; C, 0.92 nmol; D, 5.77 nmol; E, 0.92 nmol; F, 4.82 nmol. Flow rate: 0.8 ml/min.

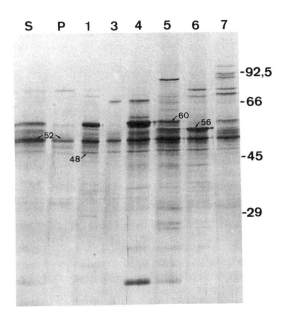

Figure 4.   SDS-PAGE of Mono Q-fractions. Sample for fractionation: Lubrol-solubilized liver microsomes of dexamethasone-treated female rats; P450-content: 2 nmol. S: Sample. P: Pass-through fraction; 1 - 7: Peaks as in Figure 2. Apparent molecular weights (kilodalton) are assigned to individual bands.

Table III. Comparative P450 determination in Mono Q fractions obtained with Lubrol-solubilized microsomes of rats treated with various inducers

Sample	P	1	2	3	4	5	6	7	Sum
PBm	36.4	5.5	3.7	7.3	5.5	5.0	2.7	0.8	70.1
DEXm	5.4	12.2	13.3	13.3	11.6	6.9	1.7	0.3	64.7
ßNFm	36.4	5.9	4.5	4.5	12.9	23.0	6.7	3.0	100.0
UTm	33.0	6.3	9.7	1.5		19.9	-	-	81.7
UTf	25.8	2.2	5.1	9.1		14.7	3.8	-	63.0

Microsomal cytochrome P450s were fractionated on Mono Q by the optimized chromatographic procedure as shown in Figure 2. Values give the percentage of applied P450 in the pooled peak fractions as determined spectroscopically by the method of Omura & Sato (72). P: pass-through. 1 - 7, peak designation as in Figure 2. PB: phenobarbital; DEX: dexamethasone; ßNF: ß-naphthoflavone; UT: untreated; m: male; f: female.

After fractionation of microsomal P450-isozymes of phenobarbital-treated rats on Mono Q, P450 2B-species can be detected by an antiserum against P450 2B1 in the pass-through fraction and the eluted fractions 1 to 6 (not shown, see Figure 2 for peak designation). Further good indications for the existence of multiple isozyme forms grouped in the subfamily P450 2B, e.g., inducible by phenobarbital (39,40), come from the following chromatographic and electrophoretic results obtained by others. Three immunoidentical P450-forms were separated on DEAE-Sepharose from liver microsomes of phenobarbital-treated rats by Sakai et al. (41). They were designated L, M and H and were distinguished by their apparent molecular weights and enzymatic activities. These P450-forms probably correspond to the isozymes RLM5, RLM6 and RLM7 which were later isolated by Backes et al. (42) and possess identical N-terminal amino acid sequences up to residue 32, indicative of 2B-isozymes. Discriminating properties are again molecular weights and enzymatic activities. Furthermore, a P450-fraction of phenobarbital-treated rats which was shown to be homogeneous by SDS-PAGE was resolved in three peaks by anion-exchange HPLC (33,34). Proof for even greater heterogeneity comes from analyses by isoelectrofocusing reported by Oertle et al. (43) who resolved a SDS-PAGE-homogeneous preparation into six bands by this technique. Further analysis demonstrated differing enzymatic activities of the fractions towards p-nitroanisole and 7-ethoxycoumarine (44). It is not clear, however, to what extent these forms are different gene products or posttranslationally modified proteins. In this context it may be important that serine phosphorylation occurs with P450 2B1/B2 (45,46) resulting in decreased enzymatic activity (47), a discriminating property of the P450-forms separated by Sakai et al. (41) and Backes et al. (42). Phosphorylated and non-phosphorylated forms may be distinguished by isoelectrofocusing and probably also by optimized HPLC-methods. Further studies will have to focus on the characterization of the isozyme forms separated by ion-exchange FPLC.

Cation-Exchange Chromatography on Mono S. The P450-components of the Mono Q pass-through fraction bind quantitatively on the cation exchange resin Mono S equilibrated in a MOPS-containing buffer at pH 7.0. Chromatographic resolution on Mono S is more sensitive to the pH than on Mono Q. Judged from the analytical runs at pH-values of 6.7, 7.0 and 7.4 highest resolution is achieved at pH 7.0 (data not detailed here). By using an optimized stepwise gradient four different fractions were obtained (Figure 5). Taken together, the combined application of anion- and cation-exchange FPLC results in separation of eleven P450-containing fractions (17). Recent

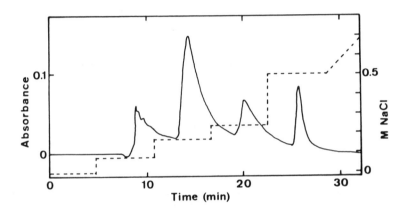

Figure 5. **Further fractionation of a Mono Q–pass–through fraction by optimized cation–exchange FPLC.** Column: Mono S HR5/5. Sample: Pass through fraction of a Mono Q–column obtained with Lubrol-solubilized liver microsomes of phenobarbital-treated male rats. Amount of applied P450: 0.83 nmol. Flow rate: 1 ml/min.

results indicate that the resolution of P450 on Mono Q can be further enhanced by the variation of the gradient profile.

Analytical Immobilized Metal Affinity Chromatography

As is evident from the elution profiles and SDS-PAGE patterns, the combined application of optimized anion- and cation-exchange FPLC reveals fractions containing more than one P450-isozyme. Further resolution of these fractions may be achieved by a subsequent chromatographic procedure based on a different separation principle. For this purpose immobilized metal affinity chromatography (IMAC) was introduced by us for P450 separation (17,48).

IMAC is based on the ability of proteins to preferentially bind to immobilized transition metal ions by their histidine and to a lesser extent by their cysteine and tryptophan residues (49). In contrast to other affinity chromatographic methods IMAC allows differential elution of structurally differing protein or peptide variants (50). It is this property of IMAC that makes it useful as an analytical separation method. This approach is reported in recent publications (51,52). Protein separations by IMAC published so far are confined to soluble proteins (e.g. 53-55). Thus, our attempt to separate microsomal P450-species by IMAC is the first application of this method to the fractionation of membrane-bound enzymes. Here, we would like to describe method optimization, show some applications and discuss the different chromatographic behaviours of soluble and membrane proteins in IMAC.

Chromatography Protocol.   Because analytical IMAC, as presented here, requires a large number of consecutive washing and equilibration steps (see below), a special FPLC-system was designed. It allows automatic control of the whole procedure resulting in reproducible elution profiles (48). The chromatographic protocol includes the following steps: (1) column wash with water; (2) charge with metal ion; (3) column wash with water; (4) column wash with the equilibration buffer; (5) column wash with the elution buffer; (6) column wash with the equilibration buffer; (7) partial metal discharge by pumping 1 ml water and a defined volume of 200 mM EDTA under reversed-flow condition; (8) column wash with equilibration buffer from the top; (9) sample injection; (10) column wash with the equilibration buffer; (11) elution from either the top or the bottom; and (12) column regeneration.

Some comments should be given to these steps. The initial and intermediate washing steps with water (steps 1,3) are necessary when phosphate is used in the equilibration and in the elution-buffers, as is done here; otherwise, metal phosphates will precipitate in the system. Column equilibration with the starting and elution buffer is necessary to wash out loosely bound metal ions. In addition, the primary aquo chelate complex is converted by ligand exchange into an eluent chelate complex. In the case of the nickel/imidazole system, this can be observed by an increase of color intensity on the column wich was also described for other metal/eluent systems (56). In order to avoid contamination of the eluted protein fractions with transition metal ions, a metal-free column section with a defined volume is produced by EDTA-washing from the bottom (step 7).

Selection of the Suitable Metal-Ion.   Adapting IMAC to a special separation problem requires the investigation of several operating parameters influencing binding and desorption of the sample. First we studied different immobilized metal ions for their ability to bind and to release spectroscopically intact P450. The following buffer has been used for equilibration (IMAC-A): 50 mM potassium phosphate, pH 7.2, 20 % glycerol, 0.2 % Lubrol PX and 0.5 M NaCl. The elution buffer (IMAC-B) contained 100 mM imidazole, in addition. Negligible P450-binding is observed with immobilized Mn^{2+}, Fe^{2+} and La^{3+}. Formation of a stable chelate on iminodiacetic acid resins was shown at least for La^{3+} (57). A different degree of binding was obtained with Cu^{2+},

Ni^{2+} and Zn^{2+} (Table IV). Copper apparently leads to very strong P450-binding as also stated by Kastner & Neubert *(58)*. In contrast, a low binding capacity is obtained with immobilized zinc. The most appropriate metal ion examined by us is Ni^{2+} (Table IV).

Table IV. Binding and recovery of cytochromes P450 fractionated by IMAC

Sample	Ion	Eluent	% of applied cytochrome P-450		
			pass-through	eluted	total
ßNFm	Zn^{2+}	NH_4Cl	56,5	≤ 1	57
ßNFm	Zn^{2+}	Imidazole	58,4	5,8	64,2
ßNFm	Ni^{2+}	NH_4Cl	15,5	12,5	28,0
ßNFm	Ni^{2+}	Imidazole	13,1	35,4	48,5
ßNFm	Ni^{2+}	Imidazole	12,3	30,1	42,4
INHm	Ni^{2+}	Imidazole	15,0	40,2	45,2
DEXm	Ni^{2+}	Imidazole	14,9	56,9	71,8
PBm	Ni^{2+}	Imidazole	21,4	47,6	68,7
PBf	Ni^{2+}	Imidazole	13,0	61,0	74,0

Samples: Lubrol-solubilized liver microsomes of rats treated with the inducers indicated (ßNF: ß-naphthoflavone, PB: phenobarbital, INH: isonicotinic acid hydrazide, DEX: dexamethasone). m and f: male and female rats resp. Chromatographic procedure as described by Roos *(48)*.

Selection of Buffer and Eluting Components. It has been shown that the presence of unprotonated electron-donor groups in the analyte is essential for protein-binding to the immobilized metal-chelate *(56)*, i.e., a slightly alkaline pH has to be selected for the equilibration and sample buffer. We have chosen a pH of 7.5 as the starting condition. Protein elution can be achieved by decreasing the pH. In the case of cytochrome P450, however, this leads to a pronounced loss of spectroscopically detectable enzymes *(48)*. Therefore, we have used imidazole as a suitable competing agent for the elution of P450. Ammonium chloride is less suitable for the elution of P450 because concentrations of up to 2 M are necessary for the complete desorption of the sample *(48)*. High salt concentrations, however, result in destabilization of cytochromes P450 *(59)* which is probably responsible for the low yields in the eluted fractions (Table IV). Recoveries of spectroscopically intact P450 with the nickel/imidazole-system are usually in the range of 42 % to 75 % of applied cytochrome (Table IV).

As a result of their optimization strategy Kastner & Neubert *(58)* selected glycine as the most appropriate reagent for the elution of P450 from Ni^{2+}-charged chelating superose. However, glycine, as a weak chelator, causes metal leakage from the column, while the metal-chelate remains stable in the presence of up to 50 mM imidazole *(51)*. Metal-chelate stability should also be considered when selecting the buffer component for equilibration and elution buffers. Tris-(hydroxymethyl)-aminomethane (Tris) strips off metal ions from the nickel-charged column; this can be observed by the accumulation of colored ions in the uncharged gel section. Another undesired effect of Tris and other organic buffers, like MES and BES, is competition with the protein for binding to the immobilized metal ion *(50)*. Therefore, we have used non-interfering potassium phosphate in the equilibration and elution buffers.

Effects of the Detergent and the Gradient Shape. Fractionation of solubilized microsomes by IMAC in the presence and absence of a detergent clearly demonstrates its importance for P450-stabilization. Using microsomes of phenobarbital-treated rats and application of a simple linear gradient, recoveries amounting to 33.9 ± 4.6 % (n = 4) of applied cytochrome P450 were obtained. In the presence of Lubrol, however, about 70 % P450 is routinely recovered by fractionation of liver microsomes of phenobarbital-treated rats. The chromatographic resolution of P450 in IMAC by the application of simple linear gradients can be dramatically increased by optimization of the gradient shape. As in ion-exchange FPLC on Mono Q (see above), combined segmented / stepwise gradients gave good results. Up to seven P450-containing fractions can be obtained by an optimized gradient (Figure 6, Table V). Analysis of these fractions by SDS-PAGE demonstrates their different protein composition (Figure 7). Because of their similar or identical molecular weights (1), identification of the individual P450-isozymes on electropherograms can only be speculative. Therefore, subsequent immunoblotting is necessary.

Table V. P450- and protein-content of fractions obtained by IMAC of solubilized microsomes

Sample: DEXm, 5.8 nmol P450.								
Fraction:	P	1	2	3	4	5	6	Sum
% P450	14.9	1.2	31.5	9.1	10.0	3.9	1.2	71.8
% Prot	26.3	≤1	34.2	8.8	5.4	13.7	2.0	90.4
[P450]	0.6	---	0.9	1.1	1.9	0.3	0.6	----

Sample: PBf, 8.4 nmol P450.								
Fraction:	P	1	2	3	4	5	6	Sum
% P450	13.0	0.3	33.5	8.7	6.8	9.9	1.8	74.0
% Prot	25.0	1.4	21.6	3.5	2.0	21.5	5.0	87.2
[P450]	0.6	0.3	1.7	2.7	3.8	0.5	0.4	----

Lubrol-solubilized microsomes of dexamethasone-treated male rats (DEXm) and phenobarbital-treated female rats (PBf) were fractionated by optimized IMAC (Ni^{2+}) as shown in Figure 6 (see also for peak designation). % P450: percent of applied P450; % protein: percent of total applied protein; [P450]: P450-content (nmol / mg protein).

Identification of Separated P450-Isozymes. Using antisera to several P450-isozymes, immunoblot analyses with microsomes of rats treated with various inducers (Table VI) allow the following conclusions. Differential elution of P450 isozymes by IMAC is evident. While P450 3A-species are predominantly found in the pass-through fraction or at low imidazole-concentrations, members of the 2B-family are mainly eluted at 40 mM imidazole concentration. Isozymes of both families are induced by phenobarbital (39,60). Low affinity to the nickel-charged gel is observed for the ethanol- or isonia-cide-inducible P450 2E1 and for cytochrome b$_5$. For all isozymes analyzed so far, antibody-reactivity is observed with more than one IMAC-fraction. As discussed above for the fractions obtained by anion-exchange FPLC, this may be explained by the presence of closely related isozyme forms in the separated fractions. Besides the discussed heterogeneity in the P450 2B-family, multiplicity is also shown for 3A-(38,61), 2A- (62) and 2H-isozymes (32). Separation of different P450 3A-forms by IMAC is evident from the peptide patterns after proteolytic digestion (Figure 8). In addition the existence of similar isozymes, expressed from different genes, posttranslational modifications such as phosphorylation (45,46) or glycosylation (63,64), might also contribute to the observed heterogeneity. Recognition and quantitation of all

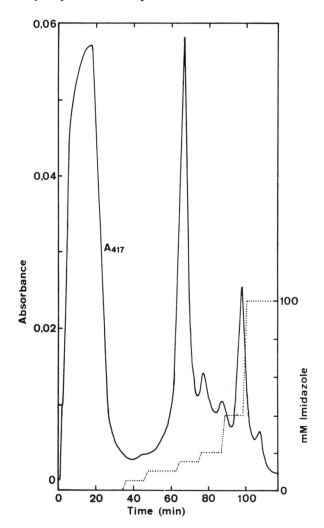

Figure 6. **Fractionation of microsomal cytochrome P450 by IMAC.** Column: Chelating Sepharose Fast Flow charged with nickel. Total bed volume: 8.2 ml, volume of the uncharged gel section: 1.7 ml. Column diameter: 1 cm. Sample: Lubrol-solubilized liver microsomes of female rats treated with dexamethasone. Amount of applied P450: 5.8 nmol. Flow rate: 1 ml/min. Solid line: Absorbance at 417 nm. Dotted line: imidazole gradient. Fractionation was done with the optimized automatic FPLC system. Buffers: IMAC-A and IMAC-B as described above (see under 'Selection of the Suitable Metal-Ion'). SDS-PAGE analyses and spectroscopically determined P450-contents of the fractions are given in Figure 7 and Table V respectively.

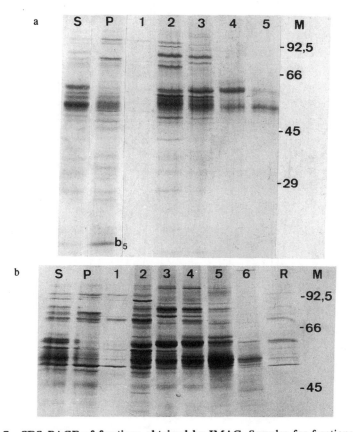

Figure 7.   SDS-PAGE of fractions obtained by IMAC. Samples for fractionation: Lubrol-solubilized liver microsomes of rats treated with (a) dexamethasone (male) and (b) phenobarbital (female); P450-content: (a) 2 nmol, (b) 8.4 nmol. Fractionation was done by the optimized method with a nickel-charged column. Peak designation as in Figure 6. S: Sample. P: Pass-through fraction. M: Molecular weight markers (kilodalton). R: Pooled residual fractions. b_5: Cytochrome b_5 (identified by immunoblotting).

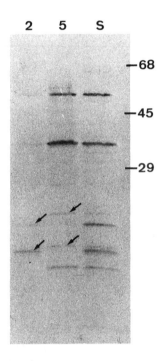

Figure 8.   Peptide patterns of IMAC-fractions after proteolytic digestion visualized after immunoblotting with antibodies against P450 3A. Sample for fractionation: Lubrol-solubilized liver microsomes of dexamethasone-treated male rats; P450-content: 10.6 nmol. Proteolytic digestion with V8-protease. S: Sample; 2, 5: peaks as in Figure 6. Position of molecular weight markers indicated (kilodalton). Arrows point to specific peptides.

these forms are important for understanding the complex interaction in the metabo-lism of drugs and other xenobiotics. Therefore, our next task is the characterization of the chromatographically separated P450-forms.

Table VI. Distribution of P450-isozymes in IMAC-fractions as analyzed by immu-noblotting after SDS-PAGE (data from Roos (48))

Peak		P	1	2	3	4	5	6
mM imidazole:		0	5	10	15	20	40	100
Sample	Antibody to							
PBm	P450 2B1/2B2	–	–	(+)	+	+	++	+
PBm	P450 3A1	++	+	++	+	+	+	+
ßNFm	P450 1A1/1A2	–	–	(+)	+	–	–	–
INHm	P450 2E1	+	+	+	–	–	–	–

Optimized IMAC was used for fractionation of solubilized liver microsomes of male (m) rats treated with phenobarbital (PB), ß-naphthoflavone (ßNF) and isonicotinic acid hydrazide (INH). Amount of applied P450 for chromatographic separation: about 1.5 nmol. Peak designation: P = pass-through, numbers 1 to 6 correspond to peak fractions as indicated in Figure 6. Symbols give relatve staining intensity on nitrocel-lulose after peroxidase reaction with chloronaphthol: -, (+), +, ++ : no visible, slight, medium, strong staining.

Chromatographic Behavior of the Separated Proteins in IMAC. Judged from the above results obtained with microsomal cytochromes P450 and cytochrome b_5, the chromatographic behavior of membrane proteins in IMAC does not seem to conform to the rules established for soluble proteins. According to Sulkowski (56), the retention of proteins on iminodiacetic acid-Me^{2+}-columns reflects the number of accessible surface histidine residues. Hemdan et al. (65) found that at least two histidines are necessary for protein binding to a nickel-charged column while for binding to a Cu^{2+}-charged gel only a single histidine residue is sufficient. Interestingly, rat cytochrome b_5 with six histidine residues does not bind to the nickel-charged gel. Its presence in the pass-through fraction is shown spectroscopically by the method of Omura & Ta-kesue (66) and by immunoblotting (unpublished results).

Judged from the three-dimensional structure of calf-liver cytochrome b_5 (67) and from sequence comparison with the orthologous rat enzyme (68), at least four histidine residues appear to be exposed at the protein surface. Among the cytochromes P450 the ethanol-inducible P450 2E1 with twelve histidines is preferentially found in the pass-through fraction. It is not known, however, to what extent these histidine resi-dues are exposed at the protein surface. On the other hand, elution of some P450-species from a nickel-charged column requires imidazole-concentrations of up to 100 mM while for soluble proteins imidazole concentrations lower than 20 mM are usually sufficient. An extraordinarily high affinity to copper-charged chelating sepharose was found, however, for the estrogen-receptor. For its elution 200 mM imidazole are ne-cessary (69). Our data suggest that interaction of the metal-chelate with surface histi-dine residues of membrane proteins can be partly abolished. Lipids or detergents may be responsible for this effect.

Concluding Remarks

The optimized chromatographic procedures presented here are part of an analytical strategy to combine the application of different methods to elaborate complex P450-patterns in microsomal samples *(70)*. Fractionation of cytochromes P450 with high-resolution chromatographic techniques in combination with immunochemical and spectroscopic methods should allow the detection and quantitation of known and unknown P450-forms with similar properties in a complex mixture of isozymes. Such comprehensive analyses of P450-profiles should help in the study of induction processes, evaluation of therapeutic measures and detection of pathological alterations of P450-patterns. The optimized methods might add alternatives to the conventional procedures for P450-purification. Scaling up is often possible *(7,17,71)*. The results obtained by IMAC of microsomal cytochromes P450 may be a first contribution to a more general study of membrane proteins separated by this method. To establish general rules, systematic studies with membrane proteins of known surface structure have to be performed.

Acknowledgements. The author thanks Mrs. Georgia Günther for expert technical assistance, Mrs. Dagmar Strotkamp for doing the peptide mapping and immunoblotting experiments, and Prof. Dr. Walter G. Hanstein for discussions on the subject and the manuscript. Thanks are also due to Pharmacia LKB GmbH (Freiburg, F.R.G.) for loan of FPLC accessories which were necessary for the automatic IMAC-system. This work was supported by the German Bundesministerium für Forschung und Technologie (FKZ 07048591).

Literature Cited

1   Ryan, D. E.; Levin, W. *Pharmac. Ther.* **1990**, *45*, 153-239.
2   Ryan, D. E.; Ramanathan, L.; Iida, S.; Thomas, P. E.; Haniu, M.; Shively, J. E.; Lieber, C. S.; Levin, W. *J. Biol. Chem.* **1985**, *260*, 6385-6393.
3   Komori, M.; Hashizume, T.; Ohi, H.; Miura, T.; Kitada, M.; Nagashima, K.; Kamataki, T. *J. Biochem.* **1988**, *104*, 912-916.
4   Kotake, A. N.; Funae Y. *Proc. Natl. Acad. Sci. USA* **1980**, *77*, 6473-6475.
5   Funae, Y.; Imaoka S. *Biochim. Biophys. Acta* **1985**, *842*, 119-132.
6   Imaoka, S.; Funae Y. *J. Biochem.* **1990**, *108*, 33-36.
7   Kastner, M.; Schulz-Schalge, T.; Neubert, D. *Toxicol. Lett.* **1989**, *45*, 261-270.
8   Gonzalez, F. J.; Nebert D. W. *Trends Genetics* **1990**, *6*, 182-186.
9   Parke, D. V.; Ioannides, C.; Lewis, D. F. V. *Can. J. Physiol. Pharmacol.* **1991**, *69*, 537-549
10 Juchau, M. R. *Life Sci.* **1990**, *47*, 2385-2394.
11 Yee, G. C.; McGuire T. R. *Clin. Pharmacokinet.* **1990**, *19*, 319-332, 400-415.
12 Sartori, E.; Delaforge, M. *Chem.-Biol. Interactions* **1990**, *73*, 297-307.
13 Wood, A. W.; Ryan, D. E.; Thomas, P. E.; Levin, W. *J. Biol. Chem.* **1983**, *258*, 8839-8847.
14 Roos, P. H.; Golub-Ciosk, B.; Strotkamp, D.; Hanstein, W.G. *Naunyn-Schmiedebergs Arch. Pharmacol.* **1991**, *344 (Suppl. 2)*, R130.
15 Thomas, P. E.; Bandiera, S.; Reik, L. M.; Meines, S. L.; Ryan, D. E.; Levin, W. *Fed. Proc. Fed. Am. Soc. Exp. Biol.* **1987**, *46*, 2563-2566.
16 Roos, P. H.; Golub-Ciosk, B.; Hanstein, W. G. In *Cytochrome P-450: Biochemistry and Biophysics;* Schuster E., Ed.; Taylor and Francis: London, 1989; pp 53-56.
17 Roos, P. H. *J. Chromatogr.* **1990**, *521*, 251-265.
18 Kuwahara, S.; Harada, N.; Yoshioka, H.; Miyata, T.; Omura T. *J. Biochem.* **1984**, *95*, 703-714.
19 Haugen, D. A.; Coon M. J. *J. Biol. Chem.* **1976**, *251*, 7929-7939.

20   Imai, Y.; Sato, R. *J. Biochem.* **1974**, *75*, 689-697.
21   Funae, Y.; Imaoka S. *Biochim. Biophys. Acta* **1987**, *926*, 349-358.
22   Imaoka, S.; Kamataki, T.; Funae, Y. *J. Biochem.* **1987**, *102*, 843-851.
23   Imaoka, S.; Terano, Y.; Funae, Y. *Biochim. Biophys. Acta* **1987**, *916*, 358-367.
24   Ichikawa, Y.; Yamano, T. *Biochim. Biophys. Acta* **1967**, *131*, 490-497.
25   Lu, A. Y. H.; Levin, W. *Biochim. Biophys. Acta* **1974**, *344*, 205-240.
26   Guengerich, F. P.; Martin M. V. *Arch. Biochem. Biophys.* **1980**, *205*, 365 - 379.
27   Dutton, D. R.; McMillen, S. K.; Parkinson, A. *J. Biochem. Toxicol.* **1988**, *3*, 131-145.
28   Kominami, S.; Hara, H.; Ogishima, T.; Takemori, S. *J. Biol. Chem.* **1984**, *259*, 2991-2999.
29   Wagner, S. L.; Dean, W. L.; Gray R. D. *J. Biol. Chem.* **1984**, *259*, 2390-2395.
30   Kaminsky, L. S.; Dunbar, D.; Guengerich, F. P.; Lee J. L. *Biochemistry* **1987**, *26*, 1276-1283.
31   Hrycay, E. G.; Gustaffson, J.-A.; Ingelmann-Sundberg, M.; Ernster L. *Eur. J. Biochem.* **1976**, *61*, 43-52.
32   Sinclair, J. F.; Wood, S.; Lambrecht, L.; Gorman, N.; Mende-Müller, L.; Smith, L.; Hunt, J.; Sinclair, P. *Biochem. J.* **1990**, *269*, 85-91.
33   Bansal, S. K.; Love, J.; Gurtoo H. L. *Biochem. Biophys. Res. Commun.* **1983**, *117*, 268-274.
34   Bansal, S. K.; Love, J. H.; Gurtoo H. L. *Eur. J. Biochem.* **1985**, *146*, 23-33.
35   Bornheim, L. M.; Correia, M. A. *Biochem. J.* **1986**, *239*, 661-669.
36   Sakaki, T.; Soga, A.; Yabusaki, Y.; Ohkawa, H. *J. Biochem.* **1984**, *96*, 117-126.
37   Kastner, M.; Schulz T. *J. Chromatogr.* **1987**, *397*, 153-163.
38   Nagata, K.; Gonzalez, F. J.; Yamazoe, Y.; Kato R. *J. Biochem.* **1990**, *107*, 718 - 725.
39   Ryan, D. E.; Thomas, P. E.; Korzeniowski, D.; Levin, W. *J. Biol. Chem.* **1979**, *254*, 1365-1374.
40   Waxman, D. J.; Walsh, C. *J. Biol. Chem.* **1982**, *257*, 10446-10457.
41   Sakai, H.; Hino, Y.; Minakami S. *Biochem. J.* **1983**, *215*, 83-89.
42   Backes, W. L.; Jansson, I.; Mole, J. E.; Gibson, G. G.; Schenkman, J. B. *Pharmacology* **1985**, *31*, 155-169.
43   Oertle, M.; Filipovic, D.; Vergères, G.; Winterhalter, K. H.; Richter, C.; Di Iorio, E. E. In *Cytochrome P-450: Biochemistry and Biophysics*, Schuster E. Ed.; Taylor and Francis: London, 1989; pp 57-60.
44   Filipovic, D.; Oertle, M.; Vergères, G.; De Pascalis, A.; Winterhalter, K. H.; Richter, C.; Di Iorio, E. E. In *Cytochrome P-450: Biochemistry and Biophysics*, Schuster E. Ed.; Taylor and Francis: London, 1989; pp 61-64.
45   Eliasson, E.; Johansson, I.; Ingelman-Sundberg, M. *Proc. Natl. Acad. Sci. USA* **1990**, *87*, 3225-3229.
46   Bartlomowicz, B.; Waxman, D. J.; Utesch, D.; Oesch, F.; Friedberg T. *Carcinogenesis* **1989**, *10*, 225-228.
47   Bartlomowicz, B.; Friedberg, T.; Utesch, D.; Molitor, E.; Platt, K.; Oesch F. *Biochem. Biophys. Res. Commun.* **1989**, *160*, 46-52.
48   Roos, P. H. *J. Chromatogr.* **1991**, *587*, 33-42.
49   Porath, E.; Carlsson, J.; Olsson, I.; Belfrage, G. *Nature* **1975**, *258*, 598-599.
50   Chicz, R. M.; Regnier, F. E. (1990) In *Guide to Protein Purification*, Deutscher M. P. Ed.; Methods in Enzymology; Academic Press: San Diego, 1990, Vol. 182; pp 392-421.
51   Belew, M.; Yip, T.-T.; Andersson, L.; Ehrnström, R. (1987) *Anal. Biochem.* **1987**, *164*, 457-465.
52   Nakagawa, Y.; Yip, T.-T.; Belew, M.; Porath, J. *Anal. Biochem.* **1988**, *168*, 75-81.
53   Heine, J. W.; de Ley, M.; van Damme, J. *Ann. N. Y. Acad. Sci.* **1980**, *350*, 364-373.

54   Kurecki, T.; Kress, L. F.; Laskowski, M. *Anal. Biochem.* 1979, *99*, 415–420.
55   Porath, J.; Olin, B. *Biochemistry* 1983, *22*, 1621–1630.
56   Sulkowski, E. *Trends Biotechnol.* 1985, *3*, 1–7.
57   Andersson, L.; Porath, J. *Anal. Biochem.* 1986, *154*, 250–254.
58   Kastner, M.; Neubert, D. *J. Chromatogr.* 1991, *587*, 43–54.
59   Imai, Y.; Sato, R. *Eur. J. Biochem.* 1967, *1*, 419–426.
60   Marie, S.; Cresteil T. *Biochim. Biophys. Acta* 1989, *1009*, 221–228.
61   Halpert, J. R. *Arch. Biochem. Biophys.* 1988, *263*, 59–68.
62   Arlotto, M. P.; Greenway, D. J.; Parkinson A. *Arch. Biochem. Biophys.* 1989, *270*, 441–457.
63   Armstrong, R. N.; Pinto-Coelho, C.; Ryan, D. E.; Thomas, P. E.; Levin W. *J. Biol. Chem.* 1983, *258*, 2106–2108.
64   Szczesna-Skorupa, E.; Kemper, B. *J. Cell Biol.* 1989, *108*, 1237–1243.
65   Hemdan, E. S.; Zhao, Y.-J.; Sulkowski, E.; Porath, J. *Proc. Natl. Acad. Sci. USA* 1989, *86*, 1811–1815.
66   Omura, T.; Takesue, S. *J. Biochem.* 1970, *67*, 249– 257.
67   Mathews, F. S.; Levine, M.; Argos, P. *J. Mol. Biol.* 1972, *64*, 449–464.
68   Ozols, J.; Heinemann, F. S. *Biochim. Biophys. Acta* 1982, *704*, 163–173.
69   Hutchens, T. W.; Yip, T.-T. *Anal. Biochem.* 1990, *191*, 160–168.
70   Roos, P. H.; Golub-Ciosk, B.; Kallweit, P.; Hanstein, W. G. *Biol. Chem. Hoppe-Seyler* 1990, *371*, 747–748.
71   Kastner, M.; Neubert, D. *J. Chromatogr.* 1991, *587*, 117–126.
72   Omura, T.; Sato, R. *J. Biol. Chem.* 1964, *239*, 2370–2378.

RECEIVED December 15, 1992

Chapter 10

Monosaccharide Compositional Analysis of *Haemophilus influenzae* Type b Conjugate Vaccine

Method for In-Process Analysis

Charlotte C. Yu Ip and William J. Miller

Department of Cellular and Molecular Biology, Merck Sharp & Dohme Research Laboratories, West Point, PA 19486

A method has been developed to analyze the monosaccharide composition of a *Haemophilus influenzae* type b vaccine designed to immunize infants against diseases caused by *Haemophilus* organisms. This vaccine consists of the *Haemophilus influenzae* type b capsular polysaccharide, polyribosylribitol phosphate, covalently coupled to an outer membrane protein complex from *Neisseria meningitidis* serogroup B. Samples of polyribosylribitol phosphate-outer membrane protein complex conjugate vaccine were hydrolyzed in trifluoroacetic acid and subjected to high-pH anion-exchange chromatography followed by pulsed amperometric detection. The vaccine was found to contain ribitol and ribose in an approximate molar ratio of 1:1, as in the structure of polyribosylribitol phosphate. In addition, the data of investigations revealed the presence of glucosamine, galactose, glucose, 2-keto-3-deoxyoctonic acid and sialic acid in approximate molar ratios of 6:5:4:2:1, which corresponds to sugar components of lipopolysaccharides integral to the structure of outer membrane protein complex. This method also can be applied to generate in process analytical data for providing the opportunity to improve process control and optimization.

The bacterium *Haemophilus influenzae* type b (Hib) is the leading cause of bacterial meningitis in infants and young children in the United States and most other developed countries. In children, more than 95% of all invasive diseases attributable to *Haemophilus* species, which include septicemia, pneumonia, epiglottitis, cellulitis, arthritis, osteomyelitis and pericarditis, are caused by this organism. Approximately 75% of all systemic Hib infections occur in children younger than 18 months of age, and the peak incidence in North America of Hib disease occurs in children 8-12 months of age. Therefore, it has been important to develop a vaccine which confers protection in children younger than 18 months old.

0097–6156/93/0529–0132$06.00/0

The major protective antigen of Hib is the external capsular polysaccharide, polyribosylribitol phosphate (PRP) (*1*). A vaccine that consisted of PRP first was used in 1974 in a field trial in Finland and found to induce protective antibody (*2*). The poor immunogenicity of PRP in children under 18 months of age (*3*) has been observed in a large number of studies (*2-6*). This is due to the "T-independent" nature of this polysaccharide antigen. Consequently, several laboratories have produced new vaccines having PRP conjugated to a carrier protein that could stimulate T cells and enlist their help in activating B cells to produce antibodies in infants. A conjugate vaccine that links PRP to an outer membrane protein complex (OMPC) from *Neisseria menigitidis* serogroup B (NMB) (*7*) (PedvaxHIB, Merck Sharp & Dohme), has been shown to induce a high rate of protection against invasive disease caused by Hib in infants as young as 6 weeks old (*8*).

The present report provides monosaccharide compositional analysis of the PRP-OMPC conjugate vaccine using a method based on high-pH anion-exchange chromatography and pulsed amperometric detection. As reported by Hardy *et al.* (*9*) the monosaccharide composition can be conveniently determined at high sensitivity by high-pH anion-exchange chromatography with pulsed amperometric detection after acid hydrolysis. This technology enables excellent resolution of most of the biologically important bacterial monosaccharides (*10*). Protocols for the resolution and the analysis of neutral and acidic oligosaccharides (*11-15*) and monosaccharides (*10,11,16-18*) have been established. In the present studies, the procedure of acid-hydrolyzed biological samples has been used for rapid analysis to provide means to monitor the purification processes for both PRP and OMPC before they are conjugated to become the final vaccine product. The benefits of in-process analytical data are numerous and almost always lead to enhanced yield and control.

Results and Discussion

The Hib conjugate vaccine, PRP-OMPC, involves multiple synthetic steps that include the bromoacetylation of PRP and the thiolation of OMPC via N-acetylhomocysteine thiolactone and the covalent coupling of these two macromolecules under mildly basic conditions (*7*). Unequivocal proof of covalent bonding can be demonstrated after degradation of the conjugate by acid hydrolysis and the analysis of the cleavage products for the presence of S-carboxymethyl-homocysteine. A gold-particle immunoelectron micrograph of a sample of PRP-OMPC that has been reacted with antibody to PRP and gold particle secondary antibody (*19*) shows that the conjugate consists of OMPC particles ranging from 40 to 100 nm in diameter with PRP covalently attached to its surface (Figure 1).

Monosaccharide Compositional Analysis Of OMPC Across Six Process Steps.
OMPC is a liposome containing polypeptides and lipopolysaccharides (LPS) with estimated molecular weight of 3-5 x 10^7 daltons and diameter of 40-100 nm.

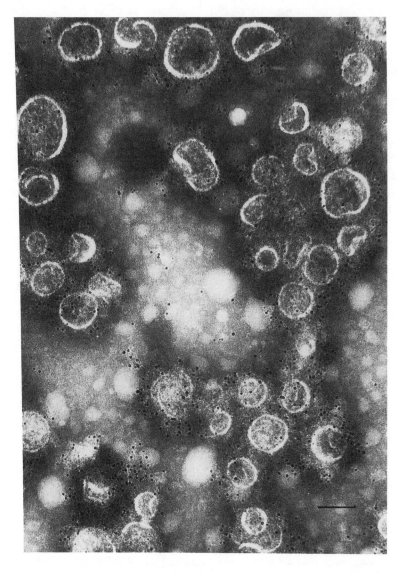

Figure 1. Immunoelectron micrograph of a sample of the PRP-OMPC conjugate vaccine. A line has been drawn to indicate a scale of 100 nm.

The sodium dodecyl sulfate (SDS)-polyacrylamide gel electrophoresis (PAGE) analyses of OMPC proteins (data not shown) were found to contain polypeptides of classes 1, 2 and 4 (*20*) in the molecular weight range of 44,000-47,000, 40,000-42,000 and 33,000-34,000, respectively. This protein also contains LPS (*21*) in the molecular weight range of 4-5,000 by sodium dodecyl sulfate (SDS) polyacrylamide gel electrophoresis (PAGE) followed by immunoblotting and reaction with monoclonal antibody to LPS (data not shown). In order to assure consistency in the LPS content of different OMPC batches, and given that LPS is an integral structural component of the OMPC, a monosaccharide compositional analysis was developed using high-pH anion-exchange chromatography with pulsed amperometric detection to provide specific information about the functioning of the OMPC purification process. Samples representing six different steps of the OMPC purification process were hydrolyzed in Reacti-Vials (Pierce, Rockford, IL) in 2 M trifluoroacetic acid (TFA) at 100°C for 5 h. After the completion of the hydrolysis, samples were dried and subjected to high-pH anion-exchange chromatography with pulsed amperometric detection analysis using a BioLc (Dionex Corp., Sunnyvale, CA) equipped with an anion-exchange analytical and guard columns, CarboPac PA1 (Dionex Corp., Sunnyvale, CA) as previously described (*10*). A 100-μg quantity of each sample was analyzed for neutral monosaccharides. The eluants used in the analysis were 10 mM NaOH (I), 50 mM NaOH containing 1M sodium acetate (II) and 500 mM NaOH (III). The separation was accomplished by isocratic elution using 10 mM NaOH as the separating solution for 35 min at a flow rate of 1 mL/min followed by cleaning and regeneration of the column (100% I to 100% II, 5 min; 100% II, 5 min; 100% II to 100% III, 1 min; 100% III, 29 min; 100% III to 100% I, 1 min; 100% I, 24 min) for a total cycle time of 100 min.

Figure 2 shows the chromatographic profiles of the neutral monosaccharides for samples analyzed across six OMPC purification steps (panels A-F). The peak eluting in the 2-min region represents the solvent peak. The monosaccharides found were identified as glucosamine (16.7 min), galactose (18.5 min) and glucose (20.4 min) in approximate molar ratios of 6:5:4, respectively, for the final product. The partially hydrolyzed peptides eluted in the region of 35-45 min. Small amounts of mannose also were found in the crude samples (panels A-E) which were eventually removed to undetectable level in the final product. The amount of monosaccharide found in each sample was tabulated and presented on a protein-normalized weight % basis (Table I). The data indicate a two-fold reduction of glucosamine, while the galactose and glucose remained essentially unchanged during processing.

Table I also gives the results for the analysis of sialic acid and 2-keto-3-deoxyoctonic acid (KDO) across the process as obtained by the chromatographic analysis of the hydrolysate. The presence of KDO in LPS is well known (*22*) and recently, the presence of covalently linked sialic acid in LPS was confirmed by a neuraminidase experiment (*23*). The labile nature of sialic acid and KDO required milder hydrolysis conditions to achieve quantitative recovery. It was determined that the optimal release of sialic acid and KDO can be accomplished by hydrolysis in 0.5 M TFA at 80°C for 70 min. The hydrolyzed samples were

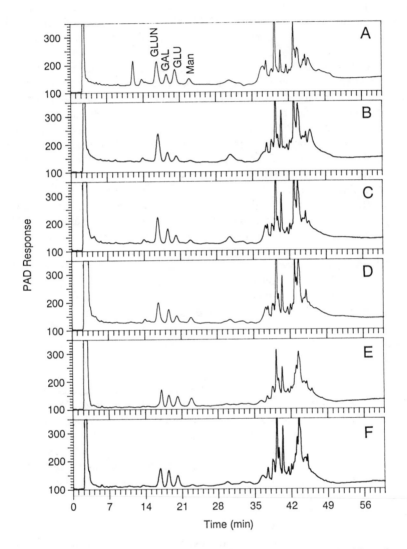

Figure 2. Chromatographic elution profiles of monosaccharides from hydrolysis of a sample of OMPC from intact cells through six purification steps (A-F) to the final product.

prepared as described previously (*10*), except that the partially hydrolyzed insoluble proteins were removed by centrifugation before the chromatographic analysis. A 150-μg quantity of each process sample was analyzed for sialic acid and KDO to obtain the chromatographic profile. The eluants were as described above for the neutral monosaccharides. The separation of sialic acid and KDO was accomplished by isocratic elution using a combination of 87% I, 5% II and 8% III for 30 min at flow rate of 1 mL/min followed by cleaning and regeneration of the column (87% I, 5% II, 8% III to 100% II, 5 min; 100% II, 5 min; 100% II to 100% III, 1 min; 100% III, 29 min; 100% III to 87% I, 5% II, 8% III, 1 min; 87% I, 5% II, 8% III, 29 min) for a total cycle time of 100 min.

Table I. Monosaccharide Composition of OMPC Across Six Process Steps

Monosaccharide (wt%)	Intact cells[1]	1st TED Extract[2]	2nd TED Extract	3rd TED Extract	4th TED Extract	Final Product
GlcNAc	0.95	1.21	1.12	1.19	0.67	0.57
Gal	0.34	0.32	0.52	0.89	0.54	0.50
Glc	0.67	0.27	0.40	0.78	0.58	0.41
Man	0.26	0.090	0.25	0.46	0.38	<0.01
NeuNAc	0.71	0.81	0.85	0.39	0.15	0.12
KDO	0.25	0.31	0.33	0.37	0.37	0.27

[1]After harvest of *N. meningitidis* serogroup B fermentation and washing with saline.
[2]Extraction in 0.1M Tris-HCl, pH 8.5, 0.01 M EDTA and 0.5% sodium deoxycholate (*25*).

Figure 3 shows the profiles of the acidic sugars for samples analyzed across the OMPC purification process. The peaks corresponding to sialic acid (14.3 min) and KDO (23.6 min) were identified as shown. Small amounts of partially hydrolyzed neutral polysaccharides eluted in the 3 to 5 min region. Partially hydrolyzed proteins eluted in the 34 min region. The data indicate that there is an approximately seven-fold reduction of sialic acid (Table I), while KDO remains essentially unchanged from crude samples to the final product. The final product was found to contain KDO and sialic acid in approximate molar ratios of 2:1, respectively.

A sample of LPS purified from OMPC was analyzed and found to contain 32.5% total carbohydrate by weight (unpublished data). The % LPS in the

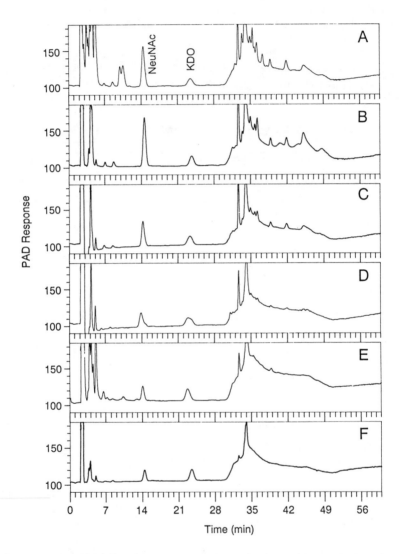

Figure 3.   Chromatographic elution profiles of acidic monosaccharides from hydrolysis of samples of OMPC from intact cells through six purification steps (A-F) to the final product.

OMPC final product was found to be 6% as estimated by dividing the total carbohydrates found in OMPC final product (Table I) by 32.5%. This is in good agreement with that estimated using the silver staining following SDS-PAGE method described previously (*24*). This result suggests that the quantitative determination of all monosaccharides in OMPC may be used as an alternative method for estimating the LPS content in OMPC.

Monosaccharide Compositional Analysis of PRP Across Four Steps. Monosaccharide compositional analysis for PRP was carried out after acid hydrolysis for four process points to provide information about the functioning of the PRP purification process. Samples representing four different stages of the PRP purification process were hydrolyzed in 2 M TFA at 80°C for 2 h. for optimal release and recovery of ribose and ribitol (*10*), monosaccharide components of PRP. After completion of the hydrolysis, the samples were dried, and 5-μg quantities of each sample were analyzed as described above for the neutral monosaccharide compositional analysis of OMPC.

Figure 4 shows the monosaccharide profiles of PRP across four process points. Ribitol and ribose eluted at 3.9 min and 30.3 min, respectively. Small amounts of glucose (20.4 min) were also found in the samples representing process points one to three (see Figure 4, panels A, B and C) and eventually removed in the final product (Figure 4, panel D). The approximate molar ratio of ribitol to ribose was found to be 1:1 in the final PRP product, consistent with the chemical composition of PRP.

Monosaccharide Compositional Analysis of PRP-OMPC conjugate vaccine. Four lots of PRP-OMPC conjugate vaccine were analyzed for PRP content via monosaccharide compositional analysis. The samples were hydrolyzed in 0.5 M TFA at 80°C for 16 h. to best accommodate the optimal release and the recovery of all monosaccharides. Panel A in Figure 5 depicts a typical HPAEC-PAD profile of neutral monosaccharides in a 50-μL sample of hydrolyzed PRP-OMPC. The monosaccharides found were ribitol, glucosamine, galactose, glucose and ribose identified as shown. The peak eluting between 2 and 3 min represents the solvent peak. The ribitol and ribose were found to have an approximate molar ratio of 1:1 and are components of PRP as evidenced by chromatographic analysis of a PRP hydrolysate (Figure 5, panel B). The presence of glucosamine, galactose and glucose (Figure 5, panel A), which were contributed by OMPC, can be confirmed by the chromatographic analysis of an OMPC hydrolysate (Figure 5 panel C). These monosaccharides are components of LPS as shown by the monosaccharide profile of a sample of LPS hydrolysate isolated from NMB (Figure 5, panel D). The data for neutral monosaccharides of LPS are in agreement with a previous investigation on the chemical composition of NMB LPS (*22*). The monosaccharide compositional analysis of PRP-OMPC aqueous bulk vaccine gives information about both the structural integrity of PRP and the LPS content in the PRP-OMPC vaccine.

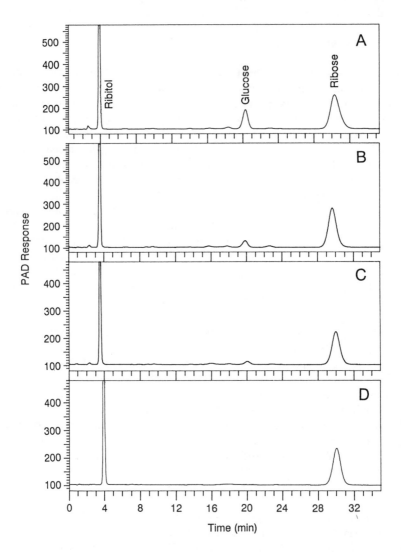

Figure 4.   Chromatographic elution profiles of monosaccharides from hydrolysis of samples of PRP from unpurified through four purification steps (A-D) to the final product.

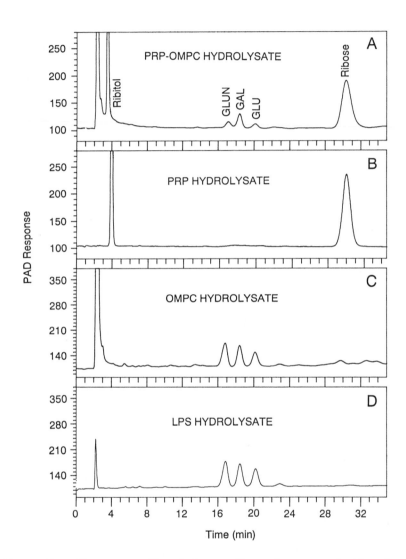

Figure 5.   Chromatographic elution profiles of monosaccharides from hydrolysis of a sample of the following: A=PRP-OMPC conjugate aqueous bulk; B=PRP; C=OMPC; D=LPS.

Conclusions

A method for monosaccharide compositional analysis of the PRP-OMPC conjugate vaccine using high-pH anion-exchange chromatography with pulsed amperometric detection was developed for providing quantitative data about the integrity of PRP in the vaccine. It was found that PRP-OMPC contained ribitol and ribose in an approximate molar ratio of 1:1, consistent with the chemical composition of PRP. A quantitative determination of LPS in the vaccine also can be achieved by this method. In addition, the monosaccharide data generated by this method provide useful in process analytical monitoring for OMPC and PRP, aiding in the understanding of how these processes function which can lead to improved process control and optimization.

Acknowledgments

We thank Ronald W. Ellis for his critical review and helpful suggestions in this work; Bohdan Wolanski for the electron microscopy; Arpi Hagopian, Ed Scattergood and members of the Merck Manufacturing Division for providing the samples for these analyses.

Literature Cited

1. Zamenhof, S., Leidy, G., Fitzgerald, P. L., Alexander, H. E. and Chargaff, E. *Am. J. Dis. CH* **1953**, *145*, 742-745.
2. Peltola, H., Kayhty, H., Sivonen, A. and Markela, H. *Pediatrics* **1977**, *60*, 730-737.
3. Daum, R. S. Granoff, D. M. *Pediatr. Infect. Dis.* **1986**, *4*, 355-357.
4. Anderson, P. Peter, G., Johnston, Jr. R. B., Wetterlow, L. H. and Smith, D. H. *J. Clin. Invest.* **1972**, *51*, 39-44.
5. Smith, D. H., Peter, G., Ingram, D. L., Harding, B. A., and Anderson, P. *Pediatrics* **1973**, *52*, 637-644.
6. Robins, J. B., Parke, J. C. Schneerson, R. and Whisnant, J. K. *Pediatr. Res.* **1973**, *7*, 103-110.
7. Marburg, S., Jorg, D., Tolman, R. L., Arison, B., McCauley, J., Kniskern, P. K., Hagopian A. and Vella, P. P. *J. Am. Chem. Soc.* **1986**, *108*, 5282-5267.
8. Santosham, M.; Wolff, M.; Reid, R., Hohenboken, M. Bateman, M.; Goepp, J.; Cortese, M.; Sack, D.; Hill, J.; Newcomer, W., Capriotti, L., Smith, J. Owen, M., Gahagan, S.; Hu, D.; Kling, R.; Lukacs, L.; Ellis, R. W.; Vella, P. P.; Calandra, G.; Matthews, H.; Ahonkhai, V. *N. Engl. J. Med.* **1991**, *324*, 1767-1772.
9. Hardy, M. R., Townsend, R. R. and Lee, Y. C. *Anal. Biochem.* **1988**, *170*, 54-62.
10. Yu Ip, C., Manam, V., Hepler, R. and Hennessey, Jr., J. *Anal. Biochem.* **1992**, *201*, 343-349.
11. Lee, Y.C. *Anal. Biochem.* **1990**, *189*, 151-162.

12. Hardy, M. R. and Townsend, R. R. *Proc. Natl. Acad. Sci. USA* **1988**, *85*, 3289-3293.
13. Wang, W. and Zopf, D. *Carbohydr. Res.* **1989**, *198*, 1-11.
14. Koizumi, K., Kubota, Y., Tanimoto, T. and Okada, Y. *Chromatogr.* **1989**, *464*, 365-373.
15. Wang, W. T., Erlansson, K., Lindh, F., Lundgren, T. and Zopf, D. *Anal. Biochem.* **1990**, *190*, 182-187.
16. Manzi, A., Diaz, S. and Varki, A. *Anal. Biochem.* **1990**, *188*, 20-32.
17. Whitfield, D. M., Stojkovski, S., Pang, H., Baptista, J. and Sarkar, B. *Anal. Biochem.* **1991**, *194*, 259-267.
18. Townsend, R. R., Hardy, M. R., Hindsgaul, O. and Lee, Y. C. *Anal. Biochem.* **1988**, *174*, 459-470.
19. Mandrell, R. E., Kim, J.J., John, C. M., Gibson, B. W., Sugai, J. V., Apicela, M. A., Grifiss, J. M. and Yamasaki, R. *J. Bacteriol.* **1991**, *173*, 2823-2832.
20. Frasch, C. E., Zollinger, W. D. and Poolman, J. T. *Rev. Infect. Dis.* **1985**, *7*, 504-510.
21. Tsai, C-M., Boykins, R. and Frasch, C. E. *J. Bacteriol.* **1983**, *155*, 498-504.
22. Limjuco, G. A., Karkhanis, Y. D., Zeltner, J. Y., Maigetter, R. Z., King, J. J. and Carlo, D. J. *J. Gen. Microbiol.* **1978**, *104*, 187-191.
23. Mandrell, R. E., Kim, J. J., John, C. M., Gibson, B. W., Sugai, J. V., Apricella, M. A., Griffiss, J. M. and Yamaski, R. *J. Bacteriol.* **1991**, *173*, 2823-2832.
24. Tsai, C-M., *J. Biol. Stand.* **1986**, *14*, 25-33.
25. Helting, T. B. Guthohrlein, G., Blackkolb, F. and Ronneberger, J. *Acta path. microbiol. scand. Sect. C.* **1981**, *89*, 69-78.

RECEIVED October 14, 1992

ADVANCES IN COLUMN ENGINEERING

Chapter 11

Zirconium Oxide Based Supports for Biochromatographic Applications

P. W. Carr, J. A. Blackwell[1], T. P. Weber[1], W. A. Schafer[2], and M. P. Rigney[3]

Chemistry Department and Institute for Advanced Studies in Bioprocess Technology, University of Minnesota, Minneapolis, MN 55455

Porous microparticulate zirconium oxide is an excellent alternative to silica-based and functionalized organic resin-based supports for the separation of biomolecules. The complex surface chemistry of this chemically and physically stable support can be exploited in a number of ways to yield biocompatible surfaces with unique selectivities. Treatment of the particles with phosphoric acid converts the amphoteric metal oxide surface to an efficient cation exchanger. When using the bare particles, the Lewis acid-base interactions between the surface zirconium(IV) ions and Lewis base solutes can be controlled by the use of competing eluent Lewis bases to give a novel ligand exchange support.

The Ideal Support. It is clear that chromatography will play a major role in the downstream processing of proteins produced by modern bioprocess techniques. One of the major limitations in process scale protein chromatography is the need for a separation medium which fulfills the requirements of an ideal support. Some of these optimal characteristics are summarized in Table I. For the purposes of high performance liquid chromatography, these supports must be mechanically stable under high pressure. Thus far, silica and very highly crosslinked organic polymers have been able to fulfill this requirement (1).

Chemical stability of the support material under a wide variety of conditions is as important as the physical characteristics of the support. In addition, the ideal support must be able to tolerate extremes of pH and temperature. Although silica is the most widely used chromatographic support, it is well known that silica and bonded phase silica are not stable outside the pH range of 2 to 8 (2,3). Above pH 8, the silica support is subject to alkaline hydrolysis. Below pH 2, the

[1]Current address: 3M Company, 270–4S–02, 3M Center, St. Paul, MN 55144
[2]Current address: Merck Sharp & Dohme, R80Y–115, Rahway, NJ 12345
[3]Current address: Ecolab, 370 North Wabasha Street, St. Paul, MN 55102

siloxane-carbon linkages in bonded phase silicas hydrolyze and slowly leach into the eluent stream (*4*). Both the dissolution and decomposition of the bonded phase are accelerated by increased temperature.

Hydrolysis of the support is a particularly severe limitation in the context of downstream processing of proteins because the use of hot alkaline media is a routine sanitization procedure in the production of biopharmaceuticals. An additional problem is the contamination of the protein effluent with undesired siliceous and organic materials. In reversed phase preparative liquid chromatography, instability of the support and the bonded ligand network also leads to contamination of the product with dissolved silica and/or ligand (*5,6*). Thus it is very difficult to maintain sterile conditions in large scale chromatographic separations using conventional supports.

Other important characteristics of an ideal chromatographic support are the availability of media in a variety of pore and particle diameters and the existence of a large fraction of mesopores (>60 A, <1000A) which will allow the rapid transport of solutes to the chromatographically active surface (*1*). The ideal support's surfaces should also be amenable to chemical modification to allow the formation of a variety of different chromatographic supports such as reversed-phase, hydrophobic interaction and ion exchange materials. Finally, the surface should be as energetically homogeneous as possible to maximize efficiency. As we have found, porous microparticulate zirconium oxide fulfills many of the requirements of the ideal support for bioseparations.

Bare Zirconium Oxide Supports. Over the past several years, we have been investigating the chromatographic properties of specially formulated porous particles of zirconium oxide (*7-23*). These particles have a nominal particle diameter of five microns with 300 angstrom average pore diameter. The surface area is nominally 30 square meters per gram. The surface of zirconium oxide, like other metal oxides, is very complex. The major surface species have been spectroscopically identified as being Bronsted acid sites, Bronsted base sites and Lewis acid sites (*12,15 and references contained therein*). These species are shown schematically in Figure 1. Bronsted acid sites arise from the acidity of surface bound hydroxyl groups and tightly bound water molecules. Bronsted base sites arise from species such as bridging oxygen atoms. Although Bronsted base sites may also be classified as Lewis base sites, no Bronsted base activity has been observed in either liquid-solid or gas-solid adsorption studies.

The Lewis acid sites, which are not found on silica, arise in a different manner than do the Bronsted sites. In the bulk of zirconium oxide, the bonding of metal and oxygen atoms is governed by the coordination geometry of the zirconium(IV) ion. Tetravalent zirconium(IV) ion is heptacoordinate in monoclinic zirconium oxide, the form of zirconia used in these studies (*15*). In the interior of the material, the coordinatively bound oxygen atoms are in resonance with the covalently bound oxygen atoms. As a result, an extensive network of bonding occurs which gives zirconium oxide its high mechanical strength.

When this bonding continuity is interrupted, at a surface for example, the full coordination of zirconium ions by oxygen atoms is not possible. As a result,

Table I. Properties of an Ideal Chromatographic Support[a]

	Silica	Zirconia	Polymeric Phases
Mechanical Stability	+ +	+ +	+
High Surface Area	+ +	+ +	+ +
Control of Average Pore Diameter	+ +	+ +	+
Control of Particle Diameter	+ +	+ +	+ +
Good Solute Mass Transfer	+ +	+ +	+
Chemically Flexible	+ +	+ +	+
Energetically Homogeneous	-	-	+
Chemical Stability	-	+ +	+ +
Thermal Stability	+	+ +	-

a. + + denotes very good performance
   + denotes good performance
   - denotes fair performance

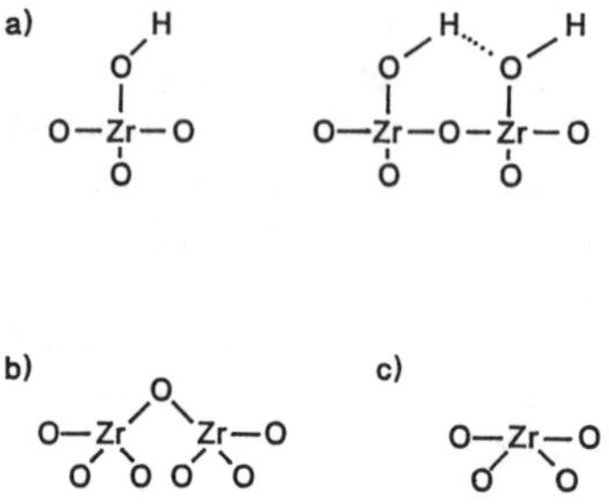

Figure 1. Schematic depicting surface site types on monoclinic zirconium oxide. (Reproduced from reference 21. Copyright 1992 American Chemical Society.)

a number of coordination sites are exposed at the surface.   These Lewis acid sites accept electron pairs from a variety of Lewis bases, depending upon the composition of the solution in equilibrium with the surface.   These coordinated Lewis bases may be water molecules, hydroxide ions, Lewis base solutes or other types of Lewis bases added to the eluent.

Despite the apparent complexity of the zirconia surface, the material is quite stable towards chemical attack.   Likewise, the bonding structure makes zirconia a very physically stable support for use in process and analytical scale chromatography.   These desirable properties have prompted others to use zirconia as a cladding to improve the physical and chemical stability of silica particles (*24*). Unfortunately, clad materials do not show the same stability as does homogeneous bulk zirconium oxide.

The stability of zirconium oxide toward extremes of pH is illustrated in Figure 2. Here, the stability of zirconium oxide was compared to chromatographic grade alumina, which is much more alkaline stable than silica, in a "static" stability study.   In this work, the particles were placed in a closed container and challenged with a variety of buffers at different pHs.   These conditions are far less aggressive than those used in a chromatographic stability test where the buffer is continuously passed over the surface of the particles.   Under continuous flow conditions, it is impossible for the buffer solution to become saturated and possibly suppress further dissolution of the support.

As shown in Figure 2, alumina dissolves in aqueous buffer at pHs greater than 12 and less than 3. In contrast, we could not detect zirconium at any pH from 1 to 14 using inductively coupled plasma spectroscopy as the detection method (*8,12*). The detection limit by this technique is 0.03 micrograms per milliliter. It should be noted that alumina is far less soluble than silica in aqueous media at all pHs (*12*).

The major challenge in using zirconia is that the unmodified surface is not biocompatible. This is analogous to the situation with silica where proteins are not readily separated on the surface of native silica gels.   The gradient elution separation of bovine serum albumin from myoglobin, while possible, produces poor results (*8,12*), as shown in Figure 3. Indeed, replicate runs on the same column show a continuous deterioration in the performance of the column because a very large fraction of the protein which was injected does not elute. The main difficulty in using porous zirconium oxide for protein separations is the need to overcome the extremely strong adsorption of proteins on the native surface.

Phosphated Zirconium Oxide. Because of its unique selectivity, there has been a great deal of interest in the use of calcium hydroxyapatite as a high performance liquid chromatographic support for the separation of proteins (*25-32*).   Great advances have been made in understanding the complex adsorption process involved in protein chromatography on calcium hydroxyapatite (25-27,33-36). Despite impressive improvements in the physical stability and particle geometry (37-40), the hydroxyapatite materials are still chemically rather unstable. We felt that it might be possible to induce the formation of a layer of zirconium phosphate on zirconium oxide, and thereby emulate some of the properties of calcium hydroxyapatite while retaining the physical and chemical stability of the underlying zirconia material.

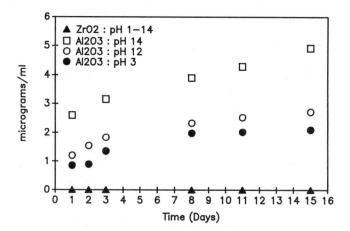

Figure 2.   Solubility of zirconium oxide versus aluminum oxide in static pH stability studies.

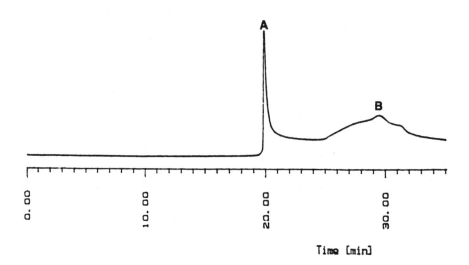

Figure 3.   Protein separation on zirconium oxide.   A = bovine serum albumin; B = myoglobin.   Gradient was from 50mM phosphoric acid to 50mM sodium phosphate at pH 10 in 10 minutes.   Flow rate was 1 mL/min at 25°C.

Depending upon chemical conditions and thermal treatment, zirconium can combine with phosphate to form a wide variety of zirconium phosphate forms *(41,42)*. As described elsewhere *(10,11)*, we have been able to modify the surface of zirconium oxide by treating it with hot, concentrated phosphoric acid solutions. Such treatment results in a surface where a significant fraction of the surface is covered with some form of zirconium phosphate and the rest of the surface is covered with a layer of adsorbed phosphate. This chemistry is illustrated in Figure 4.

After treatment of zirconia with hot phosphoric acid, the material becomes a very useful chromatographic support for the separation of proteins, as shown in Figure 5. It should be pointed out that the retention times on this material are very reproducible, as are the shapes of the chromatographic peaks *(10,11,14)*. Further studies have shown nearly quantitative recoveries of proteins with no denaturation.

The mechanism of retention of proteins on phosphated zirconia supports was examined, in part, by the experiments presented in Figure 6. In this work *(11,14)*, five proteins (cytochrome C, ribonuclease A, α-chymotrypsin, lysozyme and myoglobin) were studied. In all cases, a minimum of 0.15M pH 6 potassium hydrogen phosphate was present in the eluent. The capacity factors for the various proteins are plotted versus the total potassium ion concentration of the eluent. In the case of the open symbols, the concentration of potassium phosphate was varied. In the case of the closed symbols, potassium phosphate concentration was fixed and additional potassium was added via the use of potassium chloride. Under these conditions, it is only the total potassium concentration which controls the capacity factor. The concentration of chloride or phosphate ions does not affect retention, as long as the phosphate ion concentration remains above approximately 50mM. If the buffer phosphate concentration falls below this threshhold concentration, loss of efficiency and recovery begin to occur. These data indicate that the phosphated zirconium oxide phase is functioning as a pure cation exchange material.

This conclusion is supported by the data shown in Table II. In this work, the phosphate concentration was fixed at 200mM and the solution pH was held reasonably constant between 6.85 and 6.90. The cation in the mobile phase was varied from potassium to sodium to ammonium ion. Clearly, the capacity factors of the various proteins changed significantly with the nature of the cation. It should be further noted that only very basic proteins (e.g. lysozyme and cytochrome C) are retained on the phosphated zirconia phase. This is in marked contrast to the performance of calcium hydroxyapatite, wherein both anionic and cationic proteins may be retained. Thus, the phosphated zirconia surface does not truly emulate the behavior of hydroxyapatite.

This does not mean that the phosphated zirconia material is not a chromatographically useful material. First, the sequence of protein elution, as shown in Figure 5, is markedly different from the elution sequence of cationic proteins on conventional silica-based bonded phase cation exchangers. Second, it is not difficult to find practical circumstances under which the ability of this material to adsorb positively charged proteins is useful. One such separation is

Simplified Model of Zirconia Surface

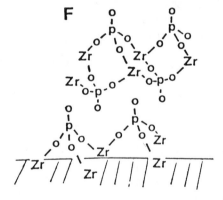

Figure 4. Proposed models for the surface species on various phosphate treated zirconias. A) terminal hydroxide group; B) bridging hydroxide group; C) Lewis acid site; D) physi-adsorbed phosphate group; E) esterified phosphate group covalently bound to the surface; F) multi-layer zirconium phosphate region resulting from partial dissolution of zirconia matrix.

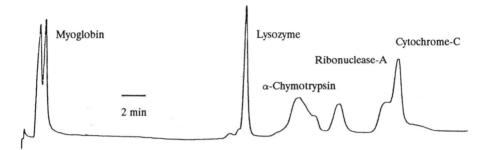

Figure 5. Chromatographic separation of cationic proteins on phosphate modified zirconia. Linear gradient elution from 50 to 500mM potassium phosphate buffer at pH 7.0. Flow rate was 0.5 mL/min.

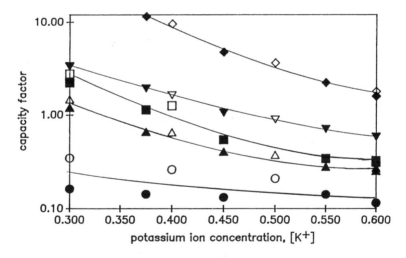

Figure 6. Isocratic protein retention as a function of potassium ion concentration. Proteins: (○) myoglobin; (▲) lysozyme; (□) α-chymotrypsin; (▽) ribonuclease A; (◇) cytochrome C. Mobile phase: (closed symbols) 0.15M potassium phosphate pH 6.0 with added potassium chloride to give indicated potassium ion concentration; (open symbols) potassium phosphate buffer at pH 6.0 at the indicated concentration.

shown in Figure 7. Immunoglobulin G can be produced in animal cell tissue culture. Such cultures require the presence of albumin and transferrin in the growth media. As shown in Figure 7, bovine serum albumin and transferrin are very easily separated from IgG on the phosphated zirconia support. The separation is excellent and it is possible to heavily overload the column before the IgG peak can no longer be resolved from the transferrin peak.

Zirconium Oxide Adsorption Mechanism Studies. In order to carry out the separation of proteins on bare zirconium oxide supports, it is essential to understand the surface chemistry responsible for the adsorption of proteins on this material. As shown in Figure 1, we believe that the Lewis acid chemistry of surface zirconium ions with certain types of solute Lewis bases is involved in the "irreversible" adsorption process for proteins. In order to gain a greater understanding of the role of Lewis acid-base processes on zirconium oxide, we carried out a series of studies using a wide variety of low molecular weight carboxylic acids on zirconium oxide.

The result of one such study is presented in Figure 8. Here we show the dependence of the capacity factors of a wide variety of substituted benzoic acids on the pKa of that derivative (21). This study was carried out in two types of buffers, acetate buffer and six similar aminosulfonate buffers with different pKa values. As shown, there is a reasonable correlation between the capacity factors and the pKas of the benzoate derivatives. Virtually all of the data shown in Figure 8 can be collapsed onto a single line, as shown in Figure 9 when the effect of pH on retention is factored out. Note that the open circles correspond to the set of six aminosulfonate buffers, whereas the closed circles correspond to the acetate buffer data.

The data indicate that an increase in pH decreases retention. A decrease in the pKa of the solute also decreases the capacity factor. These results are chemically quite significant. We do not believe that the decrease in capacity factor as the pH is raised is due to a change in the state of ionization of the benzoic acid derivative. In fact, over the entire pH range presented in Figures 8 and 9, the benzoic acid derivatives were all essentially completely ionized (except in acetate buffer). The decrease in capacity factor with pH is a consequence of the increased competition between hydroxyl ions (which are strong, hard Lewis bases) with the carboxylate solutes for the available zirconium ion Lewis acid sites.

The variation in retention with the pKa of the solute is also related to Lewis acid-base processes. The affinity of the benzoate anions for the proton is directly measured by its Bronsted basicity. One must keep in mind the fact that as the Bronsted acidity of a weak acid decreases, the Bronsted basicity of the conjugate base, that is the affinity for protons, increases. The correlation between capacity factor and pKa is easily reconciled when one considers that the proton is the quintessential hard Lewis acid. Similarly, zirconium(IV) is an exceptionally strong, hard Lewis acid. Since protons and zirconium(IV) ions are both hard Lewis acids, the correlation between proton affinity and zirconium ion affinity should be high. As mentioned above, the solutes are all essentially completely ionized under the chromatographic conditions employed here thus the degree of ionization does not significantly affect retention.

Table II. Effect of Various Cations on the Isocratic Elution of Proteins[a]

Displacing Cation	capacity factor		
	lysozyme	ribonuclease A	α-chymotrypsin
potassium	0.77	6.27	4.72
sodium	3.17	10.28	9.99
ammonium	0.79	eno[b]	18.97

SOURCE: Adapted from reference 14.
a. Mobile phase was 200mM phosphate buffer with counterion as listed above at pH 6.9. Column dimensions were 50 x 4.6 mm filled with 6-11 micron phosphated zirconia particles. Flow rate was 0.5 mL/min with detection at 280 nm.
b. eno = elution not observed.

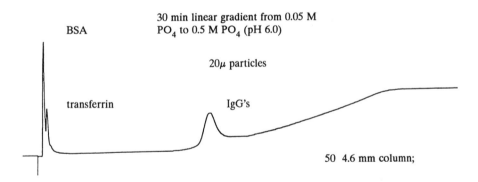

Figure 7. Separation of immunoglobulin G from a modeled fermentation broth. Linear gradient from 50 to 500mM sodium phosphate buffer at pH 6.0. Flow rate was 0.5 mL/min.

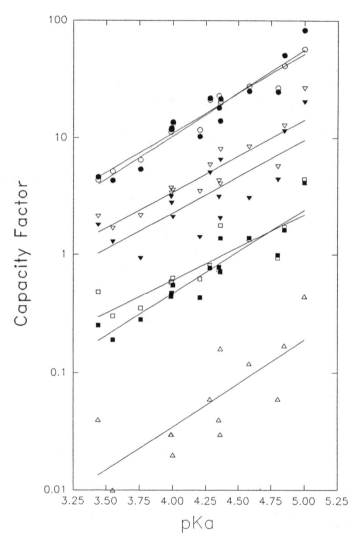

Figure 8. Plot of log k' of parabenzoic acid derivatives versus their pKa in
various buffers. (●) acetate pH 4.8; (○) PIPES pH 6.8; (▽) MOPS pH 7.2;
(▼) HEPES pH 7.5; (□) EPPS pH 8.0; (■) TAPS pH 8.4;
(▲) CHES pH 9.3; all 100mM. Solid lines indicate least squares fits.

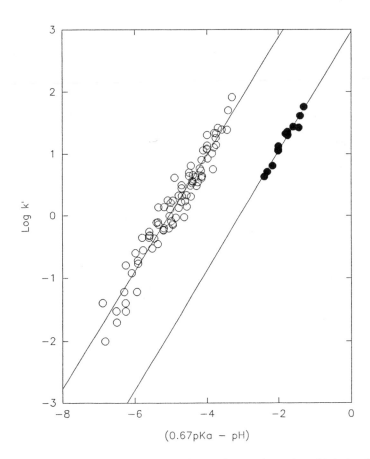

Figure 9. Correlation of the capacity factors for parabenzoic acid derivatives with solute Bronsted acidity and eluent pH. (O) aminosulfonate buffers; (●) acetate buffer. Regressions are least squares fits.

In addition to the effect of pH on retention, the eluent used in conjunction with zirconium oxide is also extremely important. As shown in Table III, the retention of benzoic acid derivatives on zirconia is very strongly influenced by the presence of Lewis bases in the eluent which may compete with solute Lewis bases for the surface Lewis acid sites. When phosphate or fluoride is added to the eluent as competing Lewis bases, there is virtually no retention of the benzoic acid derivatives. Both phosphate and fluoride are very strong, hard Lewis bases towards zirconium(IV) ions, consequently they are very strong competitors for the occupation of coordination sites on the zirconium oxide surface. A wide variety of competing Lewis bases have been evaluated as to their relative displacing strengths towards benzoic acid derivatives.

In essence, Table III constitutes an eluotropic series of displacement strength of the various Lewis bases on zirconium oxide. A rough correlation exists between the logarithmic capacity factors of the benzoic acid derivatives on the support and the logarithm of the formation constant of a complex between zirconium(IV) ions and a number of Lewis base eluents shown in this table for which data are available.

Protein Chromatography on Lewis base Modified Zirconia. A chromatogram of a set of proteins separated on the bare zirconium oxide surface when operated in a medium containing 20mM sodium fluoride is shown in Figure 10. A 0 to 0.75M sodium sulfate gradient produces a sequence of sharp, extremely well resolved peaks. This chromatogram shows that, once the Lewis acid-base chemistry of zirconium oxide is taken into account, by adding the strong Lewis base to the eluent (i.e. F⁻) it is possible to use the unmodified surface as a support for the separation of proteins. The relationship between the capacity factors of a series of proteins under similar elution conditions and the protein's isoelectric point are shown in Figure 11. Here, the filled circles represent proteins at a pH below their isoelectric point. Clearly, both cationic and anionic proteins are retained on zirconia when operated in fluoride media. This behavior is quite analogous to the behavior of calcium hydroxyapatite. We believe that the anionic proteins are mainly retained because of their ability to interact with surface zirconium(IV) ion sites via Lewis acid-base interactions. Another interesting but puzzling aspect of this material is that denatured proteins are not retained. This is entirely analogous to Gorbunov's observations with hydroxyapatite (25).

As shown in Table IV, the concentration of sodium fluoride in the eluent has a very strong effect on protein retention. As anticipated, as the fluoride concentration is increased, retention decreases considerably. Also note in this table that although the pH was 5.5, proteins with low isoelectric points, such as ovalbumin and serum albumin, are rather strongly retained. In contrast, cationic proteins, such as ribonuclease B, cytochrome C and lysozyme are less strongly retained, even in 0.5M sodium chloride media. Clearly, zirconium oxide is able to retain and separate both cationic and anionic proteins depending on the relative contributions of ligand exchange and electrostatic interactions to retention.

Table III. Eluotropic Series Based on Benzoic Acid Derivatives on Zirconia[a]
Capacity Factors at pH 6.1

Eluent Species	4-nitro-benzoate	4-cyano-benzoate	4-formyl-benzoate	4-chloro-benzoate
phosphate	0.0	0.0	0.0	0.0
fluoride	0.0	0.0	0.0	0.0
ethylphosphonate	0.2	0.2	0.2	0.4
citrate	0.3	0.2	0.2	0.6
malate	0.2	0.2	0.3	0.5
ethylenediaminetetraacetate	0.3	0.3	0.4	0.9
oxalate	0.5	0.5	0.5	0.8
succinate	0.6	0.7	0.8	1.4
glutarate	0.6	0.7	0.9	1.5
aspartate	0.7	0.6	0.7	1.5
adipate	0.7	0.7	0.9	1.7
maleate	0.7	0.7	0.7	n/a
malonate	0.9	1.0	1.1	2.0
pimelate	0.9	1.0	1.3	2.4
sulfate	1.5	1.6	1.8	3.0
glycolate	1.7	1.8	2.0	3.4
tartarate	2.9	1.4	2.2	5.8
borate	1.9	2.0	2.2	3.8
nitrilotriacetate	2.1	1.6	2.2	5.2
suberate	2.0	2.2	3.4	7.6
thiosulfate	3.0	3.0	3.4	5.9
iminodiacetate	3.1	3.1	3.9	7.7
sebacate	3.8	3.0	3.0	20.6
acetate	5.6	6.3	7.6	12.7
TRIS	7.6	6.7	6.0	15.6
formate	8.8	9.7	9.5	20.9
sulfamate	10.8	11.1	15.3	23.5
butyrate	13.0	15.6	18.7	34.4
bromide	17.0	18.2	22.9	40.4
urea	17.1	17.6	24.3	37.1
guanidine hydrochloride	20.8	22.4	27.2	eno[b]
butanesulfonate	23.5	22.8	29.6	45.2
nitrate	24.7	26.0	33.5	57.6
chloride	24.7	26.6	35.1	59.9
thiocyanate	32.3	34.3	43.3	eno
ethylene glycol	34.1	33.6	42.3	eno
thiourea	41.7	43.9	eno	eno

SOURCE: Reprinted with permission from reference 22. Copyright 1991.
a. Eluent was 20mM of the above Lewis base in 20mM MES (morpholinoethanesulfonic acid) at pH 6.1. Flow rate was 1.0 mL/min at 35°C. Injections were 10 μL volumes of 10mM solutions in water onto a 50 x 4.6 mm column.
b. eno = elution not observed.

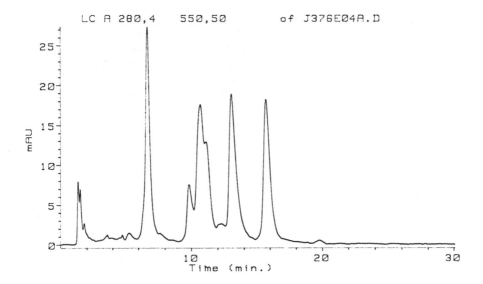

Figure 10. Protein separation on fluoride modified zirconium oxide. A linear gradient of 0-0.75M sodium sulfate in 100mM sodium fluoride and 20mM MES at pH 5.5 was used. Flow rate was 0.5 mL/min at 35°C. Protein loadings were 4.4 μg lysozyme, 15.4 μg α-chymotrypsin, 13.6 μg myoglobin and 15.4 μg cytochrome C.

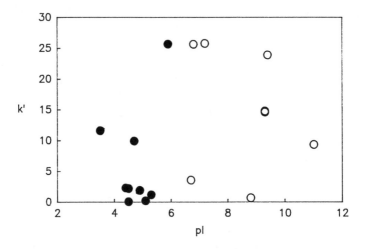

Figure 11. Correlation of capacity factor with pI for proteins on fluoride modified zirconia. Gradient elution from 0 to 0.5M sodium sulfate in 30 minutes was used. Both buffers contained 20mM sodium fluoride and 20mM MES at pH 6.2 at 35°C. Flow rate was 0.5 mL/min and detection was at 280 nm.

For process scale separations, the loading capacity and protein recoveries are critical parameters. As shown in Figure 12, significant amounts of proteins can be loaded onto the column before there is any significant change in capacity factor. The data in Table V show that the recovery of protein on zirconia in fluoride media is nearly quantitative.

We have found that when a sufficient amount of Lewis base is present in the eluent, retention times on zirconia are extremely reproducible. This amount is dependent upon the adsorption properties of the particular Lewis base on zirconia at a particular pH. By maintaining a nearly saturated isotherm for the eluent Lewis base at the surface Lewis acid sites, the solute-surface Lewis interactions are attenuated. Increases in eluent Lewis base concentration serve to further reduce the

Table IV. Protein Retention in Fluoride Media[a]

Protein	pI	Capacity Factors		
		0.50M NaF	0.10M NaF	0.02M NaF
ovalbumin	4.7	10.6	17.3	eno[b]
bovine albumin	4.7, 4.9	14.2	23.1	32.5
human albumin	4.6-5.3	17.0	25.3	34.1
apotransferrin	5.9	16.6	26.5	31.2
myoglobin	6.8, 7.3	3.5	16.7	24.6
hemoglobin	6.9-7.4	3.9	15.7	24.2
alcohol dehydrogenase	8.7-9.3	21.2	eno	eno
α-chymotrypsin	8.8	2.3	13.2	18.1
ribonuclease A	9.3	3.3	15.9	19.7
ribonuclease B	9.3	3.6	6.8	11.1
cytochrome C	9.4, 9.0	5.1	22.6	28.4
lysozyme	11.0	0.2	5.2	10.3

SOURCE: Adapted from reference 15.
a. Buffers consisted of: A) sodium fluoride and 20mM MES pH 5.5 and B) buffer A with 0.5M sodium sulfate added. Gradient was 0 to 100% B in 30 minutes then back to 0% B in 15 minutes with a 15 minute equilibration period. Flow was 1.0 mL/min at 35°C with detection at 280 and 410 nm. Typical injections were 10 μg protein in 20mM MES at pH 5.5.
b. Elution not observed.

Table V. Protein Recovery[a]

Protein	Assay #1	Assay #2
lysozyme	103.9% ± 2.5%	101.2% ± 2.2%
myoglobin	107.7% + 6.1%	95.7% + 4.1%

SOURCE: Adapted from reference 17.
a. averages of five separate recovery experiments.

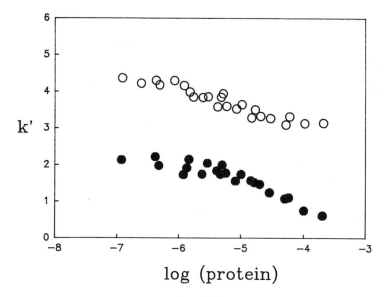

Figure 12. Loading study on fluoride modified zirconia. (○) lysozyme;
(●) lipase. Isocratic elution at 35°C and 0.5 mL/min flow rate. Eluent was
0.35M potassium chloride with 20mM sodium fluoride and 20mM TAPS at
pH 8.4.

ligand exchange contribution to retention. Decreases in the eluent Lewis base concentration decrease the ability of the eluent Lewis base to displace the bound solute Lewis base, thereby reducing recovery and efficiency.

Acknowledgments

This work was supported in part by grants to the Institute for Advanced Studies in Bioprocess Technology from the 3M Company and by grants from the National Science Foundation and the National Institutes of Health. The porous zirconium oxide particles used throughout this work were provided by the Ceramic Technology Center at the 3M Company. The authors wish to acknowledge the contributions of Dr. Eric Funkenbusch to this work.

Literature Cited

1. Unger, K.K. *Adsorbents in Column Liquid Chromatography* In *Packings and Stationary Phases in Chromatographic Techniques*; Unger, K.K., Ed.; Marcel Dekker: New York, NY, 1990; pp 331-470.
2. Hetem, M.; VanDerVen, L.; DeHaan, J.; Cramers, C.; Albert, K.; Bayer, E. *J. Chromatogr.* **1989**, *479*, 269-295.
3. Iler, R.K. *The Chemistry of Silica*; J. Wiley Interscience: New York, NY, 1979.
4. Glajch, J.L; Kirkland, J.J.;Kohler, J. *J. Chromatogr.* **1987**, *384*, 81-96.
5. Anthoni, U.; Larsen, C.; Nielsen, P.H.; Christopherson, C. *Anal. Chem.* **1987**, *59*, 2435-2438.
6. D'Ambra, A.J. Personal Communication, Department of Chemistry, University of Minneosta, Minneapolis, Minnesota.
7. Rigney, M.P.; Weber, T.P.; Carr, P.W. *J. Chromatogr.* **1989**, *484*, 273-291.
8. Rigney, M.P.; Funkenbusch, E.F.; Carr, P.W. *J. Chromatogr.* **1990**, *499*, 291-304.
9. Weber, T.P.; Carr, P.W.; Funkenbusch, E.F. *J. Chromatogr.* **1990**, *519*, 31-52.
10. Schafer, W.A.; Carr, P.W.; Funkenbusch, E.F.; Parson, K.A. *J. Chromatogr.* **1991**, *587*, 137-147.
11. Schafer, W.A.; Carr, P.W. *J. Chromatogr.* **1991**, *587*, 149160.
12. Rigney, M.P. Ph.D. Thesis, University of Minnesota, 1988.
13. Weber, T.P. Ph.D. Thesis, University of Minnesota, 1991.
14. Schafer, W.A. M.S. Thesis, University of Minnesota, 1990.
15. Blackwell, J.A. Ph.D. Thesis, University of Minnesota, 1991.
16. Blackwell, J.A.; Carr, P.W. *J. Chromatogr.* **1991**, *549*, 43- 57.
17. Blackwell, J.A.; Carr, P.W. *J. Chromatogr.* **1991**, *549*, 59- 75.
18. Blackwell, J.A.; Carr, P.W. *J. Liq. Chrom.* **1991**, *14(15)*, 2875-2889.
19. Blackwell, J.A.; Carr, P.W. *J. Liq. Chrom.* **1992**, *15(5)*, 727-751.
20. Blackwell, J.A.; Carr, P.W. *J. Liq. Chrom.* **1992**, *15(9)*, 1487-1506.
21. Blackwell, J.A.; Carr, P.W. *Anal. Chem.* **1992**, *64*, 853-862.
22. Blackwell, J.A.; Carr, P.W. *Anal. Chem.* **1992**, *64*, 863-873.
23. Blackwell, J.A.; Carr, P.W. *J. Chromatogr.* **1992**, *596*, 27- 41.

24. Stout, R.W.; DeStefano, J.J. *J. Chromatogr.* **1985**, 326, 63- 72.
25. Gorbunoff, M.J. *Anal. Biochem.* **1984**, 136, 425-432.
26. Gorbunoff, M.J. *Anal. Biochem.* **1984**, 136, 433-439.
27. Gorbunoff, M.J.; Timasheff, S.N. *Anal. Biochem.* **1984**, 136, 440-445.
28. Gorbunoff, M.J. *Meth. Enzymol.* **1985**, 117, 370-379.
29. Bernardi, G. *Meth. Enzymol.* **1973**, 27, 471-479.
30. Bernardi, G.; Giro, M.G.; Gaillard, C. *Biochim Biophys Acta* **1972**, 278, 409-420.
31. Kadoya, T.; Ogawa, T.; Kuwahara, H.; Okuyama, T. *J. Liq. Chrom.* **1988**, 11(14), 2951-2967.
32. Kadoya, T. *J. Chromatogr.* **1990**, 515, 521-525.
33. Barroug, A.; Rey, C.; Fauran, M.J.; Trombe, J.C.; Montel, G.; Bonel, G. *Bull. Soc. Chim. France* **1985**, 4, 535-539.
34. Kawasaki, T.; Niikura, M.; Kobayashi, Y. *J. Chromatogr.* **1990**, 515, 91-123.
35. Kawasaki, T.; Niikura, M.; Kobayashi, Y. *J. Chromatogr.* **1990**, 515, 125-148.
36. Inoue, S.; Ohtaki, N. *J. Chromatogr.* **1990**, 515, 193-204.
37. Kawasaki, T.; Takahashi, S.; Ikeda, K. *Eur. J. Biochem.* **1985**, 152, 361-371.
38. Hirano, H.; Nishimura, T.; Iwamura, T. *Anal. Biochem.* **1985**, 150, 228-234.
39. Kadoya, T.; Isobe, T.; Ebihara, M.; Ogawa, T.; Sumita, M.; Kuwahara, H.; Kobayashi, A.; Ishikawa, T.; Okuyama, T. *J. Liq. Chrom.* **1986**, 9(16), 3543-3557.
40. Bruno, G.; Gasparrini, F.; Misiti, D.; Arrigoni-Martelli, E.; Bronzetti, M. *J. Chromatogr.* **1990**, 504, 319-333.
41. Amphlett, C.B. *Inorganic Ion Exchangers*; Elsevier: Amsterdam, 1964.
42. Clearfield, A. *Zirconium Phosphates* In *Inorganic Ion Exchange Materials*; Clearfield, A., Ed.; CRC Press: Boca Raton, FL, 1982.

RECEIVED October 30, 1992

Chapter 12

Preparative Reversed-Phase Chromatography of Proteins

Geoffrey B. Cox

Prochrom Inc., 5622 West 73rd Street, Indianapolis, IN 46278

The mass-overloaded separations of model proteins under preparative gradient-elution conditions are characterized by strong solute-solute displacements. These displacements dominate the course of the chromatography, allowing much higher throughput, purity and mass recovery than would be *a priori* predicted. The purity and recovery of the components were not dependent upon the gradient slope. The slope does, however, influence the run time; steeper gradients allow higher production rates. The loading conditions were found to be important since too high a concentration of organic modifier prevented complete uptake of the solutes. The column efficiency was shown to be important in that a minimum number of plates (corresponding to the use of particles of 15 μm diameter or less in this case) was required to maximize the displacements. Particle sizes below 15 μm were determined to be useful only in that they allow higher mobile phase velocities and therefore higher production rates. The optimum pore diameter of the packing material is a compromise between a value large enough to allow ready access of the molecules to the interior of the particles, but small enough to give the packing material a high surface area to maximize loadability. Guidelines for the design of preparative reversed-phase gradient-elution separations of proteins are given.

Reversed-phase chromatography has been used for the analysis of peptides and smaller proteins for approximately the last 11 years (*1*),(2). It was shown that the use

0097–6156/93/0529–0165$06.00/0

of packings with large pore diameters is advantageous in protein separations. The wider pores, usually 300 or 500 Å in diameter, aid the mass transfer of the bulky proteins in and out of the particle in order fully to utilize the available surface area of the packing material and to maximize the column efficiency. Such materials have much lower surface area than the small-pore packings conventionally used for small-molecule separations. More recently, some workers have turned to non-porous particles for the analytical chromatography of proteins (3). These allow rapid separations but have very low surface area. Since the loading capacity of the column is a function of the surface area of the packing, these two approaches may not necessarily be advantageous for preparative applications.

Preparative separations of proteins and peptides have also been performed by reversed-phase chromatography over many years. Much of the early work on reversed-phase separations was directed to small-scale preparative separations. The major restraint upon the use of reversed-phase chromatography for protein purification relates to the recovery of active protein from the column. It is known (4),(5) that many proteins denature upon interaction with the highly hydrophobic reversed-phase matrix. Upon denaturation, the molecules unfold and can present a greater number of hydrophobic sites which can interact with the packing. This can lead to irreversible binding of the solute or can result in the elution of unfolded protein from the column. Many small proteins can refold and regain their activity, but, for others, inactive or partially inactivated protein is recovered from the preparative column. Thus, the application of reversed-phase techniques is restricted to that body of proteins which either do not unfold when in contact with the matrix or which re-fold once they have eluted from the column. This generally is comprised of proteins with molecular weights below 20 kD, although there are a few larger proteins which may also be purified by reversed-phase chromatography.

Theory and Modelling.

The theory of heavily mass-overloaded preparative chromatography has developed very rapidly over the past few years. This has been the result of work by the groups of Snyder (6) and Guiochon (7). Most of this work involved the isocratic separations of small molecules. Small solutes often behave in a predictable manner in that many follow the Langmuir adsorption isotherm model. It has been demonstrated that solute-solute interactions are important in preparative separations and the predictions of the displacements and tag-along effects are by now well known. It happens in practice that many solutes do not follow the competitive Langmuir model which is universally employed in the modelling, even though individually they may follow Langmuir adsorption isotherms. In such cases, the displacements can be much greater - or much less - than those predicted (8).

Much less attention has been devoted to the theory and model simulation of gradient-elution separations. For small molecules, the concept of "corresponding isocratic separations" has been introduced (9) in which the gradient separation can be predicted from model simulations relating to isocratic separations where the capacity factor of the component is set equal to the average capacity factor $(\bar{k})$ in the gradient. Other work has been performed in which the Craig countercurrent model has been applied to gradient separations (*10*).   A limitation of these approaches is that the equations and simulation programs were, as for the isocratic models, developed for solutes which follow the Langmuir adsorption isotherm and the competitive Langmuir adsorption model.   Because of the difficulties in their measurement, the adsorption isotherms of large biomolecules in reversed-phase chromatography are rarely reported.   It is, however, clear that such solutes do not (and should not) normally follow the Langmuir isotherm (*11*).   This places severe restrictions upon the application of model predictions to the preparative gradient-elution chromatography of proteins.

One feature of the modelling of preparative gradient separations is the prediction of solute - solute interactions with much the same effects upon peak shapes as seen in isocratic separations.   Given the applicability of the "corresponding" separations concept, this is not surprising.   Many differences have been found between experiment and the modelling previously carried out for isocratic separations and it was not expected that the predictions from the modelling of the gradients would be exactly followed in practice.   The practical examination of interactions between biomolecules in preparative gradient-elution chromatography is reviewed in this paper.

Experimental Studies

Choice of model compounds: A number of factors were considered when solutes were chosen for studies on the preparative reversed-phase chromatography of proteins. The solutes had to have large enough molecular weights to be representative of the proteins which may be separated by reversed-phase chromatography.   This implied a desired molecular weight somewhere between 10 000 and 20 000.   The solutes had to be readily available in reasonably large quantities and in a state sufficiently pure that no pre-purification was needed.   They also had to elute reasonably closely under most gradient conditions to allow a reasonable simulation of a "real" separation.   The final criterion was that one of the solutes had to be detectable at a unique wavelength. This was to allow the independent monitoring of its peak profile through the overlapping zones in the chromatograms, thus eliminating the need to collect and analyze multiple fractions for each experiment.

Three solutes, cytochrome c, lysozyme and ribonuclease a (RNAse), were chosen.   Cytochrome c absorbs UV radiation at a wavelength of 475 nm, where neither of the other proteins have absorption.   Under conventional reversed-phase conditions, RNAse elutes earlier than cytochrome c, whilst lysozyme elutes later. Using these three solutes with a HPLC system fitted with a diode-array detector, it was therefore possible to monitor peak shapes during the mass-overloaded separations. Use of a suitable data system allowed scaling and subtraction of chromatograms to obtain individual component envelopes.   One negative factor relating to the choice of solutes was that the three solutes differed from one another structurally and the possibility of extrapolation of the results of the studies to "real" samples may be limited.   Work designed to address this question has recently been reported [(*12*)].

Protein - Protein Displacements in Reversed-Phase Preparative LC: The first observation of strong solute - solute displacements in the reversed-phase preparative gradient-elution of proteins was reported in 1989 (*13*), during experiments designed to investigate the predictions of model simulations. These studies were carried out using a 100 nm pore C_8 bonded-phase silica, with a conventional gradient ranging from 0 to 70 % acetonitrile in a time of 20 minutes. The chromatogram arising from the introduction of a mixture containing 10 mg each of lysozyme and cytochrome c is shown in Figure 1. The solid line is the chromatogram taken at 290 nm, at which wavelength both solutes absorb radiation. The dashed line is the chromatogram taken at 475 nm, where only cytochrome c absorbs. This shows a very sharp drop in the concentration of cytochrome c at precisely the same point at which the lysozyme begins to elute. It is believed that the rapid increase in lysozyme concentration at the peak front is sufficient to displace the cytochrome c from the packing, thus affecting a very sharp separation between the two solutes. This behavior is reminiscent of the effects seen in displacement chromatography. The resemblance is more striking when the load of cytochrome c is varied whilst that of lysozyme is held at 10 mg. The resulting series of chromatograms is shown in Figure 2. These traces are all taken at 475 nm, and depict the peak shape of cytochrome c at loads varying between 0.01 mg (an "analytical" load) and 5 mg. The trailing edge of the cytochrome c peak eluted at a constant position, which coincided with the peak front of the lysozyme band. At the lowest load, the cytochrome c peak was found to have a bandwidth one-third that of a peak due to an equal load when injected alone. This narrowing of the band was due to the displacement effect. It was noted that as the sample load of cytochrome c increased, the height of the peak remained relatively constant while the band width increased with load. Such behavior is normal in displacement chromatography once a displacement train is set up and the result suggests that parallels exist between the displacements seen during elution chromatography and those obtained in "classical" displacement chromatography.

A second experiment was carried out with RNAse and cytochrome c. In this case, the cytochrome c was the later eluted peak and thus the shape of the second component could be visualised. A sharp front was observed for the cytochrome c peak, indicating that no "tag-along" effects such as are often seen in isocratic separations occurred. Subtraction of the chromatogram of cytochrome c at 475 nm (after suitable scaling) from the chromatogram taken at 290 nm confirmed that the displacement effect also occurred between these components in that the trailing edge of the RNAse peak exhibited the characteristic drop to the baseline at the point of elution of the cytochrome c peak front. The chromatograms are shown in Figure 3. Similar results have been demonstrated for a separation of cytochrome c and lysozyme where the (scaled) cytochrome c peak, as measured at 475 nm, was subtracted from the chromatogram obtained at 290 nm. The lysozyme peak, resulting as the residual chromatogram from this operation, was seen to be undistorted and to begin at the point at which the cytochrome c peak showed the onset of the displacement effect (*14*).

Effect of Gradient Slope: Modelling and simple theories of preparative gradient separations indicate that the separation should be independent of the gradient slope (*15*). This situation should be maintained in the presence of displacements;

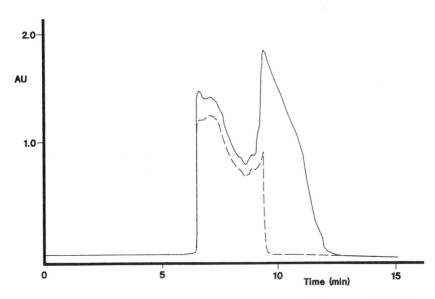

Figure 1. Chromatogram of a mixture of cytochrome c and lysozyme. Solid line, 290 nm; dashed line, 475 nm. Sample load: 10 mg each. Other conditions: Column: Zorbax PSM 1000 C8, 15 cm x 4.6 mm. 0 to 70% acetonitrile in 0.1 % trifluoroacetic acid (TFA) in 20 min. Flow rate: 1 ml/min.

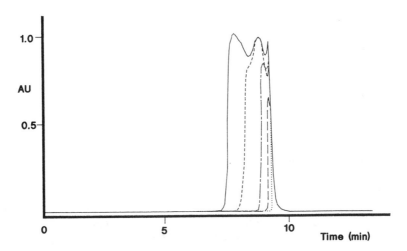

Figure 2. Overlays of 475 nm chromatograms of mixtures of cytochrome c and lysozyme. Sample loads: 10 mg lysozyme with 0.01 mg (dotted line), 0.3 mg (long dashed line), 1.0 mg (dashed / dotted line), 3.0 mg (short dashed line) and 5.0 mg (solid line) of cytochrome c. Conditions as Figure 1.

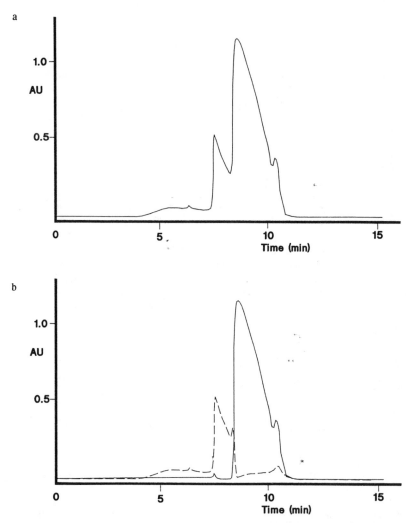

Figure 3. (a) Chromatogram of 5 mg each of cytochrome c and RNAse, detector wavelength 290 nm. (b) Overlays of the 475 nm chromatogram (solid line) and the result of subtraction of this from the 290 nm chromatogram (dashed line). Conditions as Figure 1.

provided they remain strong, there should be no effect of slope. Experiments were carried out to demonstrate this by using four gradients of different slopes [14]. The gradient details are shown in Table I. The recoveries for cytochrome c at a purity of 99.5% were calculated from the resulting chromatograms; the results are shown in Table II. These values are sufficiently close to indicate that the gradient slope had no influence upon the isolation of the product from the separation.

Table I.

Chromatographic Conditions for Gradient Slope Experiments

Column:	Zorbax Pro-10 Protein Plus, 150 x 4.6 mm.
Mobile Phase:	
	Solvent A: 0.1 % aqueous trifluoroacetic acid.
	Solvent B: 0.1 % trifluoroacetic acid in acetonitrile.
Flow Rate:	1.0 ml/min.
Gradients:	
	Range: 5 to 70% B
	Duration: (a) 15 minutes
	(b) 30 minutes
	(c) 45 minutes
(d) 60 minutes	
Detection:	UV, wavelengths 295 and 480 nm.
Sample:	Cytochrome c: 3 mg. Lysozyme: 10 mg.

Table II.

Purity and recovery of cytochrome c

Gradient duration	Purity	Recovery
15	99.46	93.1
30	99.47	93.7
45	99.50	91.3
60	99.50	95.9

Other conditions: See Table I.

Effect of Gradient Range: It is customary in the analysis of proteins to reduce the analysis times by using a narrow gradient range that just brackets the compounds of interest (16). The initial solvent composition is set to a point just below that at which the solute begins to migrate and the gradient is run at the desired rate of increase until the component is eluted, at which point a regeneration or re-equilibration step is used.

The situation in preparative chromatography is a little different. When the sample is injected at a high initial solvent strength, there is incomplete retention of the solute (14). This situation is seen in Figure 4, where a sample of cytochrome C and lysozyme (3 and 10 mg respectively) was loaded at an initial solvent concentration of 31% acetonitrile. Much of the sample was not retained, although the displacements were still seen between the retained portions of the solutes. This is indicated by the fact that the tail of the cytochrome c band is eluted earlier than the retention time of an analytical scale sample. Changing the gradient slope made little difference to the separation in terms of resolution or the extent of non-retention of the solutes. A chromatogram using a gradient slope of 0.93 %/min is shown for comparison in Figure 5. Although the distances between the peaks are increased, so are the bandwidths. This is additional evidence for the small effect of gradient slope on the preparative separation, and indicated that the non-retention of the solutes was not a function of gradient slope. A series of experiments were carried out in which the solutes were injected individually at a load of 10 mg for a variety of gradients (14). Calculation of the quantities of solute split between the retained and non-retained peaks are shown in Table III, along with the starting concentration of acetonitrile and the gradient slope.

Table III.

Ratio of solutes retained and unretained under various loading conditions

Gradient		Unretained		Retained	
initial %B	slope	Cyto	Lys	Cyto	Lys
31	2.8	0.8	0.34	0.2	0.86
31	1.4	0.8	0.34	0.2	0.86
31	0.9	0.8	0.34	0.2	0.86
24	2.8	0.3	0.06	0.7	0.94
17	2.8	0.0	0.0	1.0	1.0
10	2.8	0.0	0.0	1.0	1.0

Sample Load: 10 mg.
(Cyto = cytochrome c; Lys = lysozyme).
Other conditions as Table I.

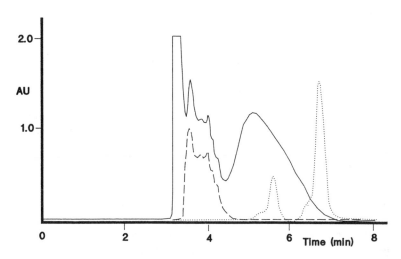

Figure 4. Chromatogram of a mixture of cytochrome c (3 mg) and lysozyme (10 mg). Solid line: 290 nm, dashed line: 480 nm, dotted line: analytical load (0.05 mg each) at 230 nm. Gradient: 31 to 45% acetonitrile in 0.1 % TFA in 5 minutes; flow rate: 1 ml/min. Column: Zorbax Pro-10 Protein Plus, 15 cm x 4.6 mm. [Figure reproduced with permission from ref 14. Copyright 1992 Elsevier Scientific Publishers.]

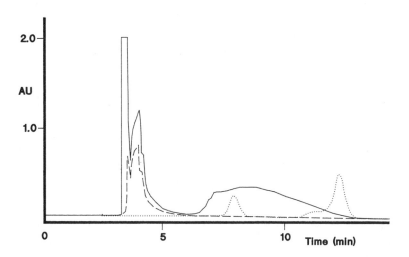

Figure 5. Chromatogram of a mixture of cytochrome c (3 mg) and lysozyme (10 mg). Solid line: 290 nm, dashed line: 480 nm. Conditions as Figure 4, except gradient duration = 20 min.

The quantity retained is much lower than the saturation capacity of the column and thus some mechanism other than complete overloading of the packing must be responsible for the incomplete retention of the solute. It is believed (17) that this is due to the displacement of adsorbed acetonitrile by the introduction of the protein solutes. Since each molecule of protein presumably displaces several molecules of acetonitrile by virtue of its larger chromatographic footprint, an introduction of a relatively large load of protein will significantly increase the local acetonitrile concentration. The retention of a protein in reversed-phase chromatography is very sensitive to small changes in solvent strength. This local increase in acetonitrile concentration is enough partially to elute the protein in the displaced solvent band as it moves through the column. A rough calculation suggests that approximately 10 moles of acetonitrile are required to be displaced by one mole of protein in order for this effect to be observed.

Reduction of the initial solvent strength increases the proportion of protein which is adsorbed. Once the solutes are introduced to the column at a low enough acetonitrile concentration that the displaced solvent no longer prematurely elutes them, then any gradient may be subsequently used. Thus, a step gradient may be used to bring the mobile phase strength to the level required for the beginning of the elution gradient, followed by a short, rapid gradient designed to elute the solutes. Figure 6 shows the result of such a process, allowing a rapid elution after loading at 10% acetonitrile. As seen in the other separations, the displacements continue to be a major feature of the separation.

Effect of pore size: Loadability in the purification of macromolecules is dependent upon the surface area available for sorption. This in turn is a function of the fraction of the surface area which lies in pores of adequate dimensions to allow access of the solute molecules. Thus, the pore diameter (and the pore size distribution) is important in determining how much sample can be loaded. Studies in ion-exchange chromatography by frontal elution (18) have indicated a relation between the kinetics of uptake of solutes with the pore diameter and the particle size.

The saturation capacities of several reversed-phase packings with different pore diameters (and hence surface areas) were determined by injection of aliquots of a standard solution of lysozyme and measurement of the quantity of the solute taken up by the column (19). From the known values of the specific surface areas of the materials, the extent of surface utilization (in terms of mg/m^2) could be calculated. These data, shown in Table IV, show a steady increase in surface utilization with pore diameter. More importantly for preparative chromatography, there is a maximum in saturation capacity at an intermediate pore diameter. It should be noted that the surface utilization of two packings with similar pore diameters but markedly differing surface areas are identical. The results demonstrate that the lysozyme molecules cannot explore the entire surface area of the packing, even when the nominal pore diameters are considerably larger than the radius of gyration of the solute. This is presumably a function of the pore size distribution, which is an important parameter in the preparative chromatogrpahy of proteins. The maximum in the saturation capacity occurs at pore diameters of around 120 Å pore diameter and this value can be taken as being close to the optimum for preparative chromatography of molecules

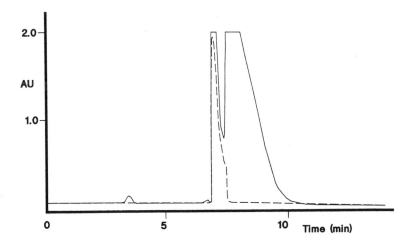

Figure 6.   Chromatogram of a mixture of cytochrome c (3 mg) and lysozyme (10 mg). Solid line: 290 nm, dashed line: 480 nm. Conditions as Figure 4, except gradient: hold at 10% acetonitrile for 2 min; step to 31 % (in 1 min); 31 to 45 % acetonitrile in 5 min.

of this and similar size. It is important to note that this pore diameter is much smaller than the 300 Å conventionally recommended for peptide and protein separations. The reduction in surface area in using large pore diameter packings is often more important than the small improvement in solute mass transfer between the phases that is given by the large pore diameters.

Table IV.

Surface utilization as a function of pore size

Pore Diameter (Å)	Surface Area (m^2/g)	Saturation Capacity (mg)	Surface Utilization [$(mg/m^2)/g$]
70	300	70	0.13
130	160	112	0.39
120	98	70	0.39
300	50	50	0.55
500	30	34	0.63
1000	20	20	0.69

Mobile phase: 5% acetonitrile in 0.1 % aqueous trifluoroacetic acid.

Effect of particle diameter: The width of the overlap zone between two adjacent solutes in the mass overloaded chromatograms of proteins appears to be a function of both the gradient parameters and the column efficiency. Experiments to vary the column efficiency while maintaining other separation parameters constant were carried out (20) using a family of C_8 250 Å pore packing materials with a range of particle diameters between 6.5 and 50 μm. The separation conditions were as described in Table V. Chromatograms of the separation of a mixture of cytochrome c and lysozyme on the 10, 20 and 50 μm packings are shown in Figure 7. The insets in the figures show the peak due to cytochrome c (i.e. 490 nm) on an expanded scale. These show the effects of band spreading on the displacement as the efficiency is decreased. Comparison of the chromatograms indicates that as the particle size is increased from 10 through 20 to 50 μm the band profile is progressively rounded until, at a particle diameter of 50 μm, the distortion in band shape due to the displacement is difficult to discern.

The separations on particles of 15 μm and below were all closely similar [20], and for this separation it is clear that there is no need for the use of extremely high efficiencies. This is evident when the recovery of cytochrome c at a purity of 99.5% is calculated. These results are shown in Table VI. The recovery is constant for particles of 15 μm and below. It begins to fall when a particle diameter between 15 and 20 μm is used; the recovery from 20 μm particles is reduced by approximately

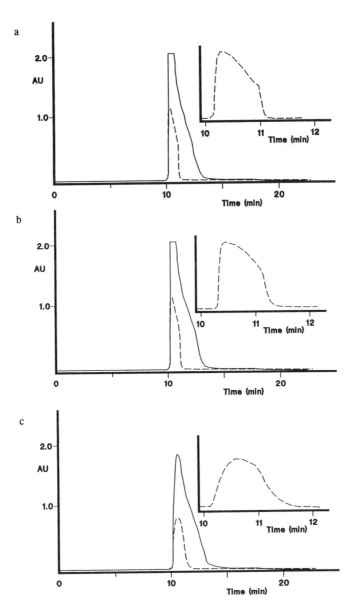

Figure 7. Chromatograms of cytochrome c (3 mg) and lysozyme (10 mg). Solid line: 300 nm, dashed line, 490 nm. Inset, expanded 490 nm chromatogram. Gradient 0 to 60 % acetonitrile in 0.1 % TFA, duration: 15 min. Flow Rate: 1 ml/min. Column: Matrex C8. Particle size: (a) 10 μm. (b) 20 μm. (c) 50 μm.

2 to 3%. Use of 50 μm particles is clearly disavantageous for the separation because of the low (75%) recovery obtained due to the extensive overlap of the two peaks.

Table V.

Chromatographic conditions for particle size study

Columns:	Matrex 250 C18, 6.5, 10, 15, 20, 50 μm. 100 x 4.6 mm
Mobile phase:	Solvents A and B, as Table 1.
Gradient:	
	Range: 0 to 60% B.
	Duration: 15 minutes.
Flow Rate:	1ml/min
Detection:	UV, 290 and 475 nm.
Sample:	Cytochrome c: 3 mg; Lysozyme: 10 mg.

Table VI.

The effect of particle diameter on recovery of cytochrome c

Particle Size (μm)	6.5	10	15	20	50
Recovery (%)	97	96	97	94	75

Conditions as Table V.

In order to achieve a high recovery in the present separation one can use a particle size of 15 μm in a 10 cm long column. Other configurations can, however, be adopted. Larger particles in a longer column could be employed to obtain the same efficiency. In this case caution must be exercised in the scale-up due to the different column configuration. Use of larger particles in the same size column but using a lower flow rate can also give the same efficiency and thus the same separation. Figure 8 shows the chromatograms arising from the use of 20 μm particles at a flow rate of 0.3 ml/min (20) with a gradient duration of 45 minutes. This results in a gradient of equal slope (in terms of % change per ml) to those in Figure 7 and any difference between the chromatograms should be attributable to the higher efficiency due to the lower flow rate. Under these conditions, the column efficiency is close to that obtained with smaller (10 to 15 μm diameter) particles. It is not surprising then, that the peak shape and recovery / purity data found in this separation are similar to those obtained with the smaller particles at higher flow rates, although the production rate is much reduced.

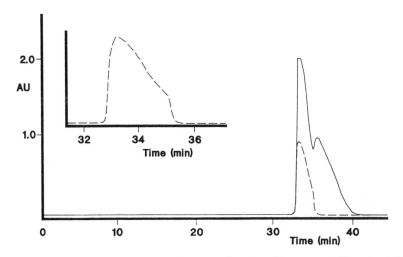

Figure 8. Chromatograms of cytochrome c (3 mg) and lysozyme (10 mg). Solid line: 300 nm, dashed line, 490 nm. Inset, expanded 490 nm chromatogram. Conditions as Figure 7, except flow rate: 0.33 ml/min, gradient duration: 45 min.

Another alternative is to use smaller particles. This will result in a column of higher efficiency, when operated under the same conditions. Since the efficiency of a column is reduced by an increase in the flow rate, it is possible to select a flow which will give the required efficiency for the protein separation. If the gradient time is adjusted to maintain a constant slope (in volume terms), the separation can then be carried out more quickly, with a corresponding increase in production rate. Because the operating pressure is a function of flow rate and particle size, there is obviously a limit to this strategy at which the pressure reaches the maximum allowed for the chromatographic equipment. This point will correspond to the highest rate of production.

Guidelines for the design of preparative elution gradients:

Sample Introduction: Protein solutes should be introduced to the column using low solvent strengths. This avoids the problems of premature elution of the solute by desorption of acetonitrile (or other organic modifier). It also allows the concentration of the sample at the head of the column, eliminating injection volume effects and allowing the injection of somewhat dilute samples. Use of too low a solvent strength should also be avoided since other work (*21*) has shown that if the packing is not wetted properly by the mobile phase the protein may not be adsorbed on the column surface and will again elute prematurely.

Gradient Slope: Since the displacements ensure that the recovery and purity of the solutes are independent of gradient slope, the gradients should be as steep and short as is practical in order to maximize the production rate. This is not true if displacements are not observed, and it is probable in such instances that too short and sharp a gradient will result in loss of resolution and, in consequence, poor purity and recovery figures. The step between loading and elution of the sample should be short to maximise production rate. A step gradient from the loading conditions to the beginning of the elution gradient is often the best choice. The starting point of the elution gradient should be close to (but preferably a little below) the concentration at which the solute becomes chromatographically mobile in order to minimize the run time. As soon as the component of interest is eluted, the regeneration procedure should be started in order to minimize the cycle time.

Pore Size: The pore diameter of the packing is important in maximizing the available surface area and therefore the column capacity for the solute. The optimum pore diameter will depend upon the conformation and molecular size of the protein, but for the range of molecular weights between 2000 and 20000, pore diameters in the range 100 to 200 Å are generally appropriate. In order for the solutes to maximize their use of the surface area, the pore size distribution should be as narrow as possible about the mean value.

Particle Diameter: The particle diameter used depends upon a number of factors. For many separations, a limited efficiency of 1000 to 2000 plates, measured for the components of interest, is sufficient to ensure narrow displacement zones and good recovery and purity of products. At conventional flow rates, particles of 15 to 20 μm are appropriate for this range of efficiency. If pressure capabilities allow, use of smaller particles in the columns will allow a higher flow rate. This means that a higher rate of production of product can be obtained.

Conclusions.

Recent work on the preparative HPLC of proteins has demonstrated that many separations are controlled by displacement effects which occur between solutes adjacent in the chromatogram. These displacements are very sharp and allow recovery of products in high purity and yield. The gradient slope and range do not affect the displacements but can be optimized to maximize the rate of production of the product. The conditions of loading the solute into the column are important, since poor selection of solvent strength at this point can limit the load which may be applied to the column.

The parameters relating to the packing material which are of importance are the pore diameter and particle size. There is an optimum pore size which allows the maximum sample load. This is often well below the 300 Å recommended for analytical separations.

The particle size is important in that it controls the column efficiency. There is a minimum efficiency below which the displacements are significantly degraded. Thus, there is a minimum particle size which is appropriate and preparative separations should be performed under conditions designed to give the required number of plates. Smaller particles at high flow rates can be used to provide a higher separation speed to enhance production rate, provided there is sufficient operating pressure available.

Literature Cited.

1. Regnier, F.E.; Gooding, K.M. *Anal Biochem.*, **1983**, *103* 1.

2. O'Hare, M.J.; Nice, E.C. *J Chromatogr.*, **1979**, *171*, 209.

3. Unger, K.K.; Jilge, G.; Kinkel, J. N.; Hearn, M.T.W. *J Chromatogr.*, **1986**, *359* (1986) 61.

4. Lau, S.Y.M.; Taneja, A.K. and Hodges, R. S. *J Chromatogr.*, **1984**, *317* 140.

5. Benedek, K.; Dong, S.; Karger, B. L. *J Chromatogr.*, **1984**, *317* 227

6. Snyder, L. R.; Cox, G. B. and Antle, P. E. *Chromatographia*, **1987**, *24, 82.*

7. Guiochon, G.; Ghodbane, S. *J Phys Chem.*, **1988**, *92*, 3682.

8. Cox, G. B.; Snyder,L. R. *J Chromatogr.*, **1989**, *483, 95.*

9. Eble, J. E.; Grob, R. E.; Antle, P. E.; Snyder, L. R. *J. Chromatogr.*, **1987**, *405*, 51.

10. Snyder, L. R.; Cox, G. B.; Antle, P. E. *J. Chromatogr.*, **1988**, *444*, 303.

11.   Cox, G. B.; Antle, P. E.; Snyder, L. R. *J. Chromatogr.*, **1988**, *444*, 1325.

12.   Cox, G. B. *J. Chromatogr.*, submitted for publication.

13.   Snyder, L. R.; Cox, G. B.; Dolan, J. W. *J Chromatogr.*, **1989** *484* 437.

14.   Cox, G. B.; Snyder, L. R. *J Chromatogr,* in press.

15.   Cox, G. B.; Antle, P. E.; Snyder, L. R. *J. Chromatogr.*, **1988**, *444*, 325.

16.   Snyder, L. R.; Stadalius, M. A. in Horváth, Cs., Ed.; High-Performance Liquid Chromatography - Advances and Perspectives, Vol 4, Academic Press, New York, 1986, p 915.

17.   Lee, A., personal communication.

18.   Kopaciewicz, W.; Fulton S.; Lee, S. Y. *J. Chromatogr.*, **1987**, *409*, 111.

19.   Cox, G. B.; Snyder, L. R.; Dolan, J. W. *J. Chromatogr.*, **1989** *,484,* 409.

20.   Cox, G. B., paper in preparation.

21.   Cox, G. B., unpublished data.

RECEIVED January 26, 1993

Author Index

Blackwell, J. A., 146
Brooks, Clayton A., 27
Burke, T. W. L., 59
Carr, P. W., 146
Cox, Geoffrey B., 165
Cramer, Steven M., 27
Frenz, John, 1
Hodges, R. S., 59
Jacobson, Jana, 77
Kalisz, Henryk M., 102
Mant, C. T., 59
Mendonca, A. J., 59

Miller, William J., 132
Nadler, T. K., 14
Prouty, Walter F., 43
Regnier, F. E., 14
Rigney, M. P., 146
Roos, P. H., 112
Schafer, W. A., 146
Schmid, Rolf D., 102
Townsend, R. Reid, 86
Weber, T. P., 146
Yu Ip, Charlotte C., 132

Affiliation Index

BioWest Research, 77
Eli Lilly and Company, 43
Genentech, Inc., 1
Gesellschaft für Biotechnologische
    Forschung, 102
Institute of Physiological Chemistry I, 112
Merck Sharp & Dohme Research
    Laboratories, 132

Prochrom Inc., 165
Purdue University, 14
Rensselaer Polytechnic Institute, 27
University of Alberta, 59
University of California—San
    Francisco, 86
University of Minnesota, 146

Subject Index

A

Acid hydrolysis, release of
    monosaccharides from glycoproteins,
    96–97
Adsorption mechanism, zirconium oxide,
    160–165
Affinity chromatography, secreted
    mammalian cell product, 44,48f,49
Amino sugars, separation, detection, and
    quantification, 101–103

Biotechnology
    challenges, 1
    chromatographic separations, 2–10
    commercial success of recombinant
        protein pharmaceuticals, 2t
    cost consciousness, 82–83
    importance of chromatography, 83
Biotechnology industry, interest in
    continuous purification systems, 14

B

Biochromatographic applications, zirconium
    oxide based supports, 152–169

C

Cation-exchange chromatography, Mono S,
    126,127f,128

183

Chemical stability
characteristic of ideal support,
152–153,154*t*
zirconium oxide supports, 155
Chromatographic concentration control,
displacement, 82–88
Chromatographic properties, zirconium
oxide supports, 153,154*f*,155,156*f*
Chromatographic resins, properties, 49–53
Chromatographic separations in
biotechnology
large-scale operations, 2–6
protein analysis by HPLC, 6–10
Chromatographic support, optimal
characteristics, 152–153,154*f*
Chromatography, advantages and
requirements, 83
Concentration control for chromatography,
displacement, 82–88
Continuous chromatography
advantages, 14
continuously stirred tank reactors, 15–16
Continuous purification of proteins, SNAP
chromatography, 14–24
Continuous selective nonadsorption
preparative chromatography
columns, 17,18*f*,19
dilutions, 19
dual-column system, 20–21,23*f*
flow rates, 20,21*t*
optimization, 19–20,22,24
packing materials, 19
selective elution, 22,23*f*
Continuously stirred tank reactors,
description, 15–16
Countercurrent flow separations,
description, 15
Cross-current flow separations,
description, 15

D

Diethylaminoethyldextran displacement,
effluent profiles, 35,36–37*f*
Displacement, description, 5*f*,6
Displacement chromatography
advantages, 27–28,83,85
chromatographic equipment, 83–84

Displacement chromatography—*Continued*
differences from elution
chromatography, 83
fraction analytical procedure, 84
histograms of separations, 85,88*f*
human insulin, 50,52*f*,53
luteinizing hormone releasing hormone,
85,87*f*
operating steps, 84
operating steps for gradient elution, 84

E

Electrochemical detection, detection
method for HPLC, 8
Enzymatic hydrolysis, release of
monosaccharides from glycoproteins, 97
Enzymes from microbial sources,
purification steps, 4*t*
Equilibrium parameters
moderately retained proteins, 31–32
strongly retained displacers, 32–33
Escherichia coli derived recombinant
product, ion-exchange chromatography,
50,51*f*

F

Fast protein chromatography, ion exchange,
microsomal cytochrome P-450 pattern
analysis, 120–128

G

Glucose oxidase
commercial importance, 108
glucose oxidation mechanism, 108
structure vs. fungal source, 108–109
Glucose oxidase isozyme separation from
Penicillium amagasakiense by
ion-exchange chromatography
compositional analysis determination
procedure, 110
crystallization, 110
electrophoretic procedure, 109
electrophoretic titration curve before
and after purification, 110,113*f*

Glucose oxidase isozyme separation from
    Penicillium amagasakiense by
    ion-exchange chromatography–*Continued*
  enzyme assay procedure, 109
  gradient slope vs. resolution, 110,112*f*
  optimal resolution conditions, 114,115*f*
  pH effect, 110,111*f*
  properties of homogeneous vs.
    heterogeneous isoforms, 114*t*
  purification procedure, 109–110
  salt effect, 110,111*f*
Glycoproteins
  characterization of carbohydrate
    structures on surfaces by HPLC, 9,10*f*
  quantitative monosaccharide analysis
    by HPLC, 92–105
Gradient elution, description, 5*f*,6
Gradient elution chromatography
  advantages, 84–85
  histograms of separations, 85,88*f*
  luteinizing hormone releasing hormone,
    85,86*f*
Gradient range, preparative reversed-phase
  chromatographic effect, 178*t*,179*f*,180,182*f*
Gradient slope, preparative reversed-phase
  chromatographic effect, 174,177*t*

H

Haemophilus influenzae type b
  importance of vaccine development, 138
  monosaccharide compositional analysis of
    conjugate vaccine, 139–147
High-performance ion-exchange
  chromatography, use for protein
  purification, 27
High-performance liquid chromatography
  protein analysis, 6–10
  quantitative monosaccharide analysis
    of glycoproteins, 92–105
Human insulin
  displacement chromatography, 50,52*f*,53
  HPLC profile of purified product, 44,45*f*
Human therapeutic pharmaceuticals,
  developments for production processes, 1

I

Ideal displacement profiles, calculation,
  33–34

Immobilized metal affinity chromatography
  description, 128
  microsomal cytochrome P-450 pattern
    analysis, 128–134
Ion-exchange chromatography
  Escherichia coli derived recombinant
    product, 50,51*f*
  secreted mammalian cell product,
    44,47*f*,49
  separation of glucose oxidase isozymes
    from *Penicillium amagasakiense*,
    108–115
Ion-exchange displacement chromatography
    of proteins
  adsorption isotherms, 35,36*f*
  apparatus, 35
  effluent profile of DEAE-dextran, 37*f*
  effluent profile of β-lactoglobulins,
    37,38*f*,39
  effluent profile of protamine, 39*f*
  equilibrium parameter determination for
    moderately retained proteins, 31–32
  equilibrium parameter determination for
    strongly retained displacers, 32–33
  ideal displacement profile calculation,
    33–34
  multicomponent equilibrium theory, 30–31
  salt effect, 27
  simulated effluent profile of
    DEAE-dextran, 35,36*f*
  simulated effluent profile of
    β-lactoglobulins, 37,38*f*,39
  simulated effluent profile of protamine,
    39,40*f*
  steric mass-action ion-exchange
    equilibrium theory, 28,29*f*,30
Ion-exchange fast protein liquid
  chromatography, microsomal cytochrome
  P-450 pattern analysis, 120–128
Isocratic elution, description, 5*f*

L

β-Lactoglobulin displacement,
  effluent profiles, 37,38*f*,39
Large-scale chromatographic purification
  of pharmaceuticals, 2–5
Lewis base modified zirconia
  capacity factor vs. isoelectric point,
    164,166*f*

Lewis base modified zirconia–*Continued*
  protein chromatography, 164,166–169
  protein loading, 167,168*f*,169
  protein retention, 164,167*t*
  protein separation, 164,166*f*
Light-scattering detection, method
  for HPLC, 8–9
Liquid chromatographic separation of
  microsomal cytochromes P-450,
  techniques, 118
Luteinizing hormone releasing hormone
  displacement chromatography, 85,87*f*
  gradient elution chromatography, 85,86*f*
  purification by preparative
    reversed-phase sample displacement
    chromatography, 66–67,70–80

M

Media of different pore and particle
  diameters, characteristic of ideal
  support, 153,154*t*
Metal affinity chromatography,
  immobilized, microsomal cytochrome
  P-450 pattern analysis, 128–134
Methanolysis, release of monosaccharides
  from glycoproteins, 93,95*f*,96
Microsomal cytochrome P-450 pattern
  analysis with fast protein liquid
  chromatography
  cation-exchange chromatography on
  Mono S, 126,127*f*,128
  detergent selection, 119
  gradient shape vs. resolution, 121,122–123*f*
  inducer effects, 121,123–125*f*,126*t*
  isozyme identification, 121,126
  optimization of resolution, 120*t*,121
  sample load vs. resolution, 121*t*,124*f*
  sample preparation for analytical
    fractionation, 119–120
  undesired effects of sample
    pretreatment, 119–120
Microsomal cytochrome P-450 pattern
  analysis with immobilized metal
  affinity chromatography
  behavior of separated proteins, 134
  buffer selection, 129
  chromatographic protocol, 128
  detergent vs. gradient shape,
    129,130*t*,131–132*f*

Microsomal cytochrome P-450 pattern
  analysis with immobilized metal
  affinity chromatography–*Continued*
  eluting component selection, 129
  isozyme identification, 130,133*f*,134*t*
  metal ion selection, 128,129*t*
Moderately retained proteins, equilibrium
  parameter determination, 31–32
Mono S, cation-exchange chromatography,
  126,127*f*,128
Monosaccharide(s)
  quantitative analysis of glycoproteins
    with HPLC, 92–105
  structures, 93,94*f*
Monosaccharide compositional analysis of
  Haemophilus influenzae type b
  conjugate vaccine
  immunoelectron micrograph of vaccine,
    139,140*f*
  outer membrane protein complex,
    139,141–145
  polyribosylribitol phosphate, 145,146*f*
  polyribosylribitol phosphate–outer
    membrane protein complex conjugate
    vaccine, 145,147*f*
Monosaccharide derivatives
  postcolumn derivatization, 100
  precolumn derivatization, 97–100
Moving belt separation, description, 15
Multicomponent equilibrium, theory, 30–31

N

Neutral sugars, separation, detection, and
  quantification, 101–103

O

On-line mass spectrometry, detection
  method for HPLC, 9
Operating modes, large-scale
  chromatographic purification of
  pharmaceuticals, 5*f*
Optimum recovery, determination, 20
Outer membrane protein complex,
  monosaccharide compositional analysis,
  139,141–145

P

Packing material, large-scale chromatographic purification of pharmaceuticals, 3–4

Particle diameter, preparative reversed-phase chromatographic effect, 181–186

Penicillium amagasakiense, separation of glucose oxidase by ion-exchange chromatography, 108–115

Peptide(s), preparative reversed-phase sample displacement chromatography, 59–80

Peptide mapping, use of HPLC, 9

pH stability, zirconium oxide supports, 155,156*f*

Polyribosylribitol phosphate monosaccharide compositional analysis, 145,146*f*

use in Hib vaccine, 139

Polyribosylribitol phosphate–outer membrane protein complex conjugate vaccine, monosaccharide compositional analysis, 145,147*f*

Pore size, preparative reversed-phase chromatographic effect, 180,181*t*

Postcolumn derivatization, monosaccharide derivatives, 100

Precolumn derivatization, monosaccharide derivatives, 97–100

Preparative elution gradient design guidelines, 186–187

Preparative reversed-phase chromatography of proteins
applications, 172
elution gradient design guidelines, 186–187
gradient range effect, 178*t*,179*f*,180,182*f*
gradient slope effect, 174,177*t*
model compound selection, 173
modeling, 172–173
particle diameter effect, 181–186
pore size effect, 180,181*t*
protein–protein displacement procedure, 173–174,175–176*f*
theory, 172–173

Preparative reversed-phase sample displacement chromatography for purification of pharmaceutically important peptides
comparison of approaches, 77,79*t*,80
low levels of organic modifier effect, 74–78
peptide elution in absence of organic modifier, 67,72–73*f*,74
peptide studied, 66–67
standard multicolumn approach, 67,70–71*f*

Process chromatography in production of recombinant products
differentiation of early from late stages in purification process, 49,51*t*
environmental impact of wastes, 57
process capacity for each step, 53,55*f*,57
process control chart for manufacturing step, 53,54*f*
product source effect on product design, 43–49
properties of chromatographic resins, 49–53
worthwhile process improvements, 53,54–56*f*,57
yield increase vs. step, 56*f*,57

Proinsulin, HPLC profile of purified product, 44,45*f*

Properties of chromatographic resins elution mode, 50,51*t*,*f*,52*f*,53
uniform size distribution of particles, 49–50

Protamine displacement, effluent profiles, 39–40*f*

Protein(s)
continuous purification by SNAP chromatography, 14–24
ion-exchange displacement chromatography, 27–40
preparative reversed-phase chromatography, 171–187

Protein analysis by high-performance liquid chromatography, 6–10

Protein chromatography, Lewis base modified zirconia, 164,166–169

Protein loading, Lewis base modified zirconia, 167,168*f*,169

Protein pharmaceutical industry, challenges, 1

Protein product source, process design effect, 43–49

Protein retention
  Lewis base modified zirconia, 164,167*t*
  zirconium oxide, 157,159*f*
Protein separation
  Lewis base modified zirconia, 164,166*f*
  zirconium oxide, 157,159*f*,160,161*f*
  zirconium oxide supports, 155,156*f*
Purification of proteins, continuous, SNAP
  chromatography, 14–24
Purification methods, influencing
  factors, 60
Purification process, differentiation of
  early from late stages, 49,51*f*

Q

Quantitative monosaccharide analysis of
  glycoproteins by high-performance
  liquid chromatography
acid hydrolysis for monosaccharide
  release, 96–97
comparison of reported monosaccharide
  compositions, 104,105*t*
detection of neutral and amino sugars,
  101–103
detection of sialic acids, 103–104
enzymatic hydrolysis for monosaccharide
  release, 97
methanolysis for monosaccharide release,
  93,95*f*,96
monosaccharide structures, 93,94–95*f*
postcolumn derivatization of
  monosaccharide derivatives, 100
precolumn derivatization of
  monosaccharide derivatives, 97–100
quantification of neutral and amino
  sugars, 101–103
quantification of sialic acids, 103–104
separation of neutral and amino sugars,
  101–103
separation of sialic acids, 103–104

R

Recombinant DNA based production methods
  origin and scale, 1
Recombinant DNA technology, role in
  development of biotechnology
  industry, 2

Recombinant products, role of process
  chromatography in production, 43–57
Recombinant protein pharmaceuticals,
  large-scale chromatographic
  operations, 2–6
Resins
  chromatographic, properties, 49–53
  selectivity effect on protein
    purification, 5
Reversed-phase chromatography
  advantages, 84
  applications, 84,171–172
Reversed-phase chromatography in gradient
  elution mode, disadvantages for
  peptide purification, 59–60

S

Salt, protein purification effect, 22,24
Sample displacement chromatography
  application to purification of
    pharmaceutically imported peptides,
    66–67,70–80
  development of multicolumn approach,
    61,66,68–69*f*
  displacement of desired hydrophilic
    peptide component by hydrophobic
    impurities, 61,64–65*f*
  displacement of hydrophilic impurities
    by desired hydrophobic peptide
    component, 61
  principles, 60–61,62–65*f*
Secreted mammalian cell product
  affinity chromatography, 44,48*f*,49
  ion-exchange chromatography,
    44,47*f*,49
Secreted protein products, heterogeneity,
  44,46*f*
Selective nonadsorption preparative
  chromatography
  adsorption to anion-exchange sorbent
    above isoelectric point, 16–17,18*f*
  continuous system, 17,18*f*,19
  dual-column system, 20–21,23*f*
  flow rates, 20,21*t*
  optimization, 19–20,22,24
  selective elution, 22,23*f*
  separation methods, 17

Sialic acids
  separation, detection, and
    quantification, 103–104
  structures, 93,95*f*
Sodium fluoride, protein retention effect,
  164,167*t*
Source of protein product, process design
  effect, 43–49
Step elution, description, 5*f*,6
Steric mass-action ion-exchange equilibrium
  protein multipoint attachment and steric
    hindrance, 28,29*f*
  theory, 28–30
Strongly retained displacers, equilibrium
  parameter determination, 32–33
Surface site types, zirconium oxide
  supports, 153,154*f*,155
Surface species models, zirconium
  oxide, 157,158*f*
Synthetic peptides
  disadvantages of purification by
    reversed-phase chromatography in
    gradient elution mode, 59–60
  factors affecting purification, 60
  preparative reversed-phase sample
    displacement chromatography, 59–80
  requirements for purification methods, 59

T

Therapeutic protein products, purification
  process effect on quality and safety, 43–49

U

Ultraviolet absorbance, detection method
  for HPLC, 8

Z

Zirconium oxide adsorption mechanism,
  capacity factors vs. pK_a, 160,162–165
Zirconium oxide supports
  chemical stability, 155
  chromatographic properties,
    153,154*f*,155,156*f*
  pH stability, 155,156*f*
  phosphate addition, 155,157–161
  protein separation, 155,156*f*
  surface site types, 153,154*f*,155

Production: C. Buzzell-Martin
Indexing: Deborah H. Steiner
Acquisition: Anne Wilson
Cover design: Sue Schafer

Printed and bound by Maple Press, York, PA